Routledge Revivals

Everything Has a History

In this collection, first published in 1951, the central theme is that everything has a history, and that we cannot fully understand anything without some knowledge of its history. Professor Haldane writes mainly on geology, astronomy and zoology, but includes a variety of other topics, including eugenics, Einstein, and C. S. Lewis. His outlines of zoology, of the geology of England, and of the evidence for astronomical theories will be of great use to students and teachers.

Everything Has a History

J. B. S. Haldane

Routledge
Taylor & Francis Group

First published in 1951
by George Allen & Unwin Ltd

This edition first published in 2016 by Routledge
2 Park Square, Milton Park, Abingdon, Oxon, OX14 4RN

and by Routledge
711 Third Avenue, New York, NY 10017

Routledge is an imprint of the Taylor & Francis Group, an informa business

© 1951 J. B. S. Haldane

Publisher's Note
The publisher has gone to great lengths to ensure the quality of this reprint but points out that some imperfections in the original copies may be apparent.

Disclaimer
The publisher has made every effort to trace copyright holders and welcomes correspondence from those they have been unable to contact.

A Library of Congress record exists under LC control number: 52003159

ISBN 13: 978-1-138-95486-1 (hbk)
ISBN 13: 978-1-315-66670-9 (ebk)

EVERYTHING
HAS A HISTORY

BY

J. B. S. HALDANE

LONDON
GEORGE ALLEN AND UNWIN LTD
RUSKIN HOUSE FORTY MUSEUM STREET

PRINTED IN GREAT BRITAIN
in 12 *point Bembo type*
BY C. TINLING AND CO. LTD.,
LIVERPOOL, LONDON, AND
PRESCOT

PREFACE

I SHOULD have liked to call this book *The History of England, and other essays*. Had I done so I should merely have deceived my readers. A history of England generally turns out to be the history of a few of the better known people of England in each century of the last two thousand years. I think it should mean the history of our Land, and should include what we know of the history of its people before any written records were made.

Most of these essays appeared in the *Daily Worker*, others in the *Modern Quarterly*, the *Rationalist Annual, Coal* and *British Ally*. I have to thank the King in Parliament for permission to republish them, though I have no doubt that permission would have been granted me even had I no legal right to republish.

The last essay in the book was delivered, as an evening discourse, at a Conference on Genetics, Paleontology, and Evolution held at Princeton University as part of the commemoration of its bicentenary. I should have preferred to have spoken on a more specialized and less speculative topic. But I am sensible of the honour done to me in asking me to speak to a wider audience than that of our more strictly scientific meetings. In this case I certainly have to thank Princeton University for permission to republish. The essay in question has appeared in *Genetics, Paleontology and Evolution*. But owing to the economic situation, this book will probably not have wide sales in Britain, though I cordially recommend it. I therefore asked, and obtained, leave to publish my contribution on this side of the Atlantic.

It will be seen that the central idea running through the book is that everything has a history, often a very strange one, and that most things have a future, often a very strange one too. My ideas as to this future clearly colour my account of the present, and of course they may be wrong. But the notion that the future will be like the present, or the past which we remember in our own lives, is certainly wrong.

I have tried to give this book a certain unity by confining it to astronomy, geology, and biology. So I have not included a number of articles on medicine, hygiene, politics, and so on, published in recent years. I have then proceeded to destroy this unity by adding two articles attacking Mr. C. S. Lewis. Some people will regard these articles as in bad taste. They are at least in the central tradition of English literature, and in much better taste than, for example, some of Milton's polemics. Besides which, Mr. Lewis, if he chooses to reply, has two considerable advantages over me. He is, as befits a student of literature, a better stylist than I. And he will be able to show, without serious difficulty, that I have contradicted myself repeatedly. I certainly have, because my thought (or if he prefers that expression, my prejudice) has developed, and I think some statements are false which I formerly thought were true.

I shall be criticized for trying to compress geology into five thousand words, and classificatory zoology into eight thousand. The answer is that some people may read these sketches who would not read a textbook of these subjects, and that if I have not failed completely in my intention, some who have read what I have written will go on to read textbooks. If they do they might do worse than remember that even textbooks leave a great deal out.

There is a need to present the results of science to every possible audience, at every level of education. I hope that I have done so in a way which will interest one of the many possible audiences.

CONTENTS

CONTENTS

VI. AN OUTLINE OF ZOOLOGY

VII. ZOOLOGICAL ESSAYS

VIII. HOW ANIMALS BEHAVE

IX. EVOLUTION

X. GREAT MEN

XI. CONTROVERSIAL

XII. HUMAN EVOLUTION, PAST AND FUTURE

I

THE HISTORY OF ENGLAND

I

The Beginnings

IN this series of articles I intend to give a brief sketch of the history of England and Wales. By this I mean, not the history of the peoples who have lived in these countries, but of the lands themselves. Scots readers will doubtless complain that I have left Scotland out. There is a good reason for this. The rocks of northern Scotland are much older than most of those of England and Wales. And they have been much more violently disturbed, so that we do not know their history in anything like the detail with which we know those of England.

Probably the best popular account of English geology is Professor Stamp's *Britain's Structure and Scenery*.* However, its author completely ignores the theory, which is now gaining more and more adherents, that the continents have shifted their position in geological time. Many geologists think that Britain and Ireland were once close to Newfoundland, and that the Atlantic was only formed in the last two hundred million years.

To understand the geological history of England we must know a little of world geology. The earth consists of a series of concentric layers, getting denser as we go down. The ocean floors are formed of a group of rocks of which basalt is a representative in which magnesium silicate predominates. The same layer is found under the continents, but at a depth of about 25 miles. Above it is a thick layer largely of granite, and above the granite usually sedimentary rocks, formed by the wearing down either of granite or of older sedimentary rocks. This lighter layer, in which aluminium silicate predominates, covers the continents and the shallow seas round them, for example the North Sea and English Channel, but not the deeper Bay of Biscay.

England has sometimes been above the sea and sometimes below it, but probably never either very far from land, or very deep

* Collins, 1946, 16/-.

under the sea.   It is the result of a struggle between two processes, erosion, mainly by water, which wears down mountains, and folding, which builds them up.   Now erosion goes on all the time, though naturally it is quickest where there is plenty of rain and the countryside slopes steeply.   But folding is intermittent. Since the earliest known rocks were formed there seem to have been nine or ten revolutions, as the geologists call them, in which new mountains were built.

Each revolution lasted for tens of millions of years, and was then succeeded by a quieter period usually lasting for about two hundred million years.   No one knows what caused the revolutions.   The most likely theory is that heat is generated in the rocks by radioactivity quicker than it can escape, and that consequently currents of molten rock are formed, on which the granite layer drifts and crumples like slag on molten metal. The revolutions were world-wide affairs.   The last one produced the Alps, the Himalayas, the Rocky Mountains, and the Andes, besides many smaller ranges.

Now let us begin our English history.   A little over five hundred million years ago, Britain was part of a land area with mountains and volcanoes.   At this time there were living things in the sea, but they have left very few fossils.   On the other hand there is no evidence at all of any life on land.   To judge from the first land plants found much later, it is unlikely that there were any plants larger than lichens.   There were certainly no four-footed animals, and probably no insects or snails.   Northern Scotland contains a fair number of rocks laid down in this remote time which we call pre-Cambrian; there are more in Anglesey, but very few in England and Wales.   ˙

About five hundred million years ago the land began to sink in an area stretching from Spitzbergen through Norway and Britain.   Before this time there was probably no Atlantic Ocean, and the trough in question may have been the beginning of it. In this trough sand and mud were laid down to form the rocks of what is called the Cambrian system, so called because they were first found in Wales, though much better developed in western North America.   They are of immense interest because they contain the first fossil animals of which we know much; though there are enough pre-Cambrian fossils to make it quite clear that life had been going on for a very long time before the Cambrian.

Near the shore lines, where coarse sediments were laid down, the most striking animals were creatures something like giant woodlice, called trilobites. In the deeper waters lived graptolites, strange floating colonies of polyps. There were sponges, corals, worms and molluscs, and other shellfish of various kinds. But there is no trace of anything remotely resembling a fish. Apart from the vertebrates, most of the main types of sea animals had already evolved by Cambrian times. What is more, many of them have made no very obvious progress since.

The bivalved shellfish have altered remarkably little. The modern descendants of the Cambrian echinoderms, which were mostly stalked, are sea-urchins and starfish which move about. The trilobites have been replaced by lobsters, crabs, and the like, which have more specialized limbs. Sessile animals like oysters and corals were already fairly perfected. Burrowers and crawlers were well developed. So were floaters. But the place of swimmers was probably mainly taken by animals like coiled squids, which moved by jet propulsion. This is an advanced method for moving in the air, but a very primitive one in water, being practised by such simple creatures as jellyfish, as well as the more advanced cephalopods like the squid and octopus. If there were any finned animals in the Cambrian they left no clear fossils.

The Cambrian lasted about a hundred million not very eventful years, and passed over into the Ordovician. More rocks were laid down in the British trough, but, probably as the result of rock folding, volcanoes developed in several places. They probably started under the sea, and later formed islands in it like the Lipari islands in the Mediterranean, or many of the West Indies. The volcanic rocks which they formed proved harder than most of the surrounding sediments, and often remain as mountains to-day for this reason. Among the mountains built of Ordovician volcanic rock are Snowdon, Cader Idris and Helvellyn. In their neighbourhood are beds where the sediment consists largely of volcanic ash. So we can date their eruptions fairly accurately in terms of the unfortunate trilobites which were killed by them.

For fossils are the geologists' date marks. One speaks of the *Olenellus* zone where a particular kind of trilobite was common, as one speaks of the Plantagenet or Tudor period.

There are no vertebrate fossils in the British Ordovician, but a few fragments of primitive fish are found in shore deposits of the Russian and North American Ordovician. It looks as if the first fish had evolved in fresh water, and a few bones had been swept into the sea. If so we may yet find fresh water deposits of this period, with complete skeletons of our remote ancestors. But this will not be in Britain, where the Ordovician rocks of western Wales and the Lake District were mostly formed in fairly deep water.

The Silurian rocks were laid down in the same trough. There was less volcanic activity, and sometimes the water was clear enough to yield white mud which became limestone. The animals were not so very different, except that a few fish appeared. They were very unlike any modern fish, as they had no paired fins and no jaws, but a rounded mouth like the modern lamprey. Most of them were heavily armoured, probably as a protection against clawed animals like giant scorpions which flourished at this time. They probably lived by scavenging in the mud like modern sturgeons. Some of them may have been our ancestors, but it is more likely that our ancestors lived in fresh water at this time.

The Cambrian, Ordovician and Silurian periods lasted a total of two hundred million years. They furnished most of the rocks of Wales, southern Scotland, and the Lake district, and also many of the hard rocks which lie below southern and eastern England.

The Silurian period came to an end with a revolution which built Scotland and most of Wales, which I shall next describe.

Life Reaches the Land

I HAVE just described the first two hundred million years of the geological history of England. At the end of the Silurian period the country was involved in a phase of mountain-building called the Caledonian revolution. Some mountains had already been formed in what are now the Scottish Highlands. The folding process now also spread to Lowland Scotland and Wales. A great chain of mountains was formed passing through Norway, Scotland, Wales, and Ireland, and probably going on through Newfoundland and Maine, which were then much nearer to Britain. In Britain the folds ran north-eastwards, roughly parallel to the old trough. Although the mountains formed may have been as high as the Andes or Himalayas, and great masses of old rocks were pushed over younger ones, the process was probably very slow. There were doubtless occasional severe earthquakes, but not necessarily any worse than those which occur in regions like Japan, California, or Java, where mountains are forming at present.

Even while they were forming, these mountains began to be worn down. The rocks formed at this time are called Devonian, and include the Old Red Sandstone. Just as clay is laid down when erosion is slow, and may later form shale or slate, so sand and gravel denote rapid currents and steep mountains. The English Devonian rocks were laid down in several different basins, for the old sea trough had been broken up by mountain-building. In one of these troughs we have evidence of a great disaster. For a bed of fish bones, the Ludlow bone bed, covering at least a thousand square miles of the upper Silurian, tells of the sudden death of hundreds of millions of fishes, perhaps through the drying up of a lagoon as the land rose, perhaps through the sudden entrance of the sea into a freshwater lake.

In many parts of the world we find evidence of large shallow lagoons at this time. But its greatest interest is that it shows the first evidence of life on land. The earliest well-known land plants, from Wales and Aberdeenshire, were something like mosses, though probably nearer to the club mosses or Lycopodiums. There were insects, whose remains are very badly preserved. The earliest known were wingless, but winged forms soon appeared which probably resembled grasshoppers; and towards the end of the Devonian the first bones of four-footed vertebrates are found.

During much of the Devonian period the Scottish Highlands were already above water. So was a land mass stretching from southern Ireland to Wales, northern England and southern Scotland. The Devonian rocks of Devon were laid down in fairly deep sea, those of Scotland in shallow water where plenty of coarse sediments were formed.

In the sea evolution went on fairly quickly. Some of the fish developed paired fins and jaws. This gave them much greater possibilities of movement and of choosing their food. So their brains also developed. One group of fishes had jointed bony bases to their fins, and some of these transformed their fins into legs as the lagoons dried up. The original function of legs was probably to allow fish to crawl from one waterhole to another. However, some of them took to life on land towards the end of the Devonian period. Most modern fish are derived from a different group of Devonian fish, whose fins were supported by a fan-like set of thin bones, without a bony axis. The earliest fossils of four-footed animals have been found in Canada in late Devonian rocks, though of course we do not know where the first of them originated. They were something like newts, but had a good deal bonier heads, and their limbs were probably not so good for walking. None of them seems to have got to Britain in Devonian times.

At the opening of the next great period, the Carboniferous, much of England and Lowland Scotland were under the sea, though Wales and the south Midlands were not. The mountains had mostly been worn down, and thick masses of limestone were laid down in the clear waters. After some millions of years northern Scotland and the surrounding lands seem to have been raised again, for huge quantities of coarse sediments were carried down by rivers to form the Millstone Grit beds of Yorkshire. Gradually the sea bed was raised to a flat plain subject to occasional submergence below sea level.

On these swampy plains the coal measures were laid down. Plants had evolved since Devonian times, and included ferns and several kinds of tree, but there were as yet no flowers. The coal was formed in swamps, different kinds having different origins. Thus ordinary coal was probably formed from decayed leaves and stems, but boghead coal mainly from spores. The animals of the coal swamps included amphibians, still resembling newts, though some were as big as crocodiles. The insects had already achieved flight, and as there were no birds or bats to harry them, they had the air to themselves. The rulers of the air were great dragon-flies with wings a foot long, and bodies as thick as pencils. Other insects resembled cockroaches and grasshoppers. But flies, bees, butterflies and beetles came later.

The rocks of the carboniferous period cover most of Ireland. In England they form the Pennine range, and of course occur in all coalfields. In Scotland they are found in the Midland valley, but here there was a swarm of small volcanoes, whose cores remain as such hills as the Bass Rock, North Berwick Law, and Edinburgh Castle Rock. Towards the end of the Carboniferous period another revolution began, called the Hercynian, after the Hartz mountains in Germany, or sometimes the Armorican, after Brittany. The folds run east and west, and across the Caledonian folds in South Wales and Southern Ireland. On the other side of what is now the Atlantic, they continue as the Appalachian mountains. This folding went on into the next period, the Permian, and was worldwide.

At about the same time coal was formed in North America, Europe, Siberia, and China, while there were glaciers in South Africa, Australia, Southern India, and South America. Many geologists now think that at that time the coalbearing regions lay in the tropics, the icy ones near the South Pole, which was then in South Africa, and that since then the continents have drifted to their present positions. Apart from the coal, there is plenty of other evidence that England had a tropical climate at this time. For example the fire-clay which usually forms the floor of coal seams, that is to say the soil on which the forests grew, is laterite, a substance which to-day is only being produced in hot regions.

The Hercynian revolution folded England appreciably, and lifted most of it well above sea level. Among the folds which date from this time are that of which the Mendip hills form the remains; and the Pennine range, running north and south, is thought to date from the same time, the direction of folding having been deflected by the earlier mountains of which Wales and the Lake district are remains.

There has been very little mountain formation in Britain since this date, about two hundred million years ago. That is why we have a relatively flat country. It is also why the rocks formed since this time are far less folded, and it is much easier to make out what happened.

3

The Age of Reptiles

AS a new period of mountain building put an end to the flat swamps in which the coal was laid down, sandstones were formed in the lower lying parts of Britain, just as they had been in Devonian times. They are called the new red sandstone, and are found in the western Midlands as far north as Lancashire, with a small patch in Ayrshire. Later on in the same period, the Permian, a salt lake something like the modern Caspian Sea, stretching from the Pennines to Germany, laid down the peculiar magnesian limestone of Durham, Yorkshire, and Lincolnshire.

Meanwhile living creatures were changing rapidly. There were already reptiles in Carboniferous times, but it was not until the Permian that they were well enough adapted to land life to colonize the land in great numbers. The most important difference between amphibians and reptiles is that reptiles have eggs with tough shells which can be laid anywhere, while amphibians have shell-less eggs, which must be laid in water or very damp places. Thus they are still tied to the water for part of their lives. The Permian reptiles were fairly like modern ones, but some of them could stand up on their four legs so that their bellies did not touch the ground. There were many different lines, of which one developed the differentiation of the teeth into incisors, canines, and molars, found in most living mammals. It is among them that our ancestors must be sought. However, we have no record of such animals in Britain. The greatest number of their remains have been found in South Africa.

The Permian passed over into the Triassic. Throughout this period Wales and much of England and Scotland were above water, and the rest was covered by lakes or shallow seas in which sand, pebble and marl beds were laid down over wide areas of England. Towards the end of the Trias some of these lakes became salt and dried up, forming salt beds as the Caspian Sea is doing now. These salt beds were buried under mud without dissolving,

and are now worked in Cheshire and to a less extent in Durham.
The British Triassic rocks are poor in fossils, as might be
expected of salt lakes laid down in desert surroundings. However
there are bones not only of fish but of a fair number of reptiles,
and near the top of the system, of a little shrew-like beast classified
as a mammal. Since it can only be classified on its bones we do not
know if it had warm blood or suckled its young. Nor till a fossilized
pregnant female or an egg is found, shall we know whether it bore
living young. But it was probably hairy. We are fairly sure that
some of the reptiles nearest to the ancestry of mammals had hair,
because we find pits in their upper jaw-bones similar to those to
which the muscles moving a cat's or rabbit's whiskers are attached.
And if they had whiskers they probably had hairy coats.These early
mammals did not evolve much further for nearly a hundred million
years, or at least their bones did not. But during this time they may
have been making the physiological adjustments needed for
temperature control, milk production by the female, and so on.

In the next period, the Jurassic, the English Midlands mostly
sank under the sea, and at different times sand, sandstones, clays
and corals were formed in shallow water, and limestones when the
water was deeper. These seas were very full of life, the best known
animals being ammonites, a kind of cuttlefish living in spiral shells.

But near the shoreline we find remains of a variety of reptiles,
including the monstrous dinosaurs. This was the great age of
reptiles. Dinosaurs up to 80 feet long lived on land, the largest
probably living in swamps, but some walking on their hind_legs
like modern kangaroos.

Other reptiles went back to the sea. The ichthyosaurs lived
like modern porpoises and sperm whales, and some were nearly as
large as the latter. They brought forth their young alive, as an
air breather must if it spends its whole life at sea. The plesiosaurs
were also sea reptiles with paddles, and long necks adapted for
snapping fish. They were unlike any living creatures. Finally
the pterodactyls achieved flight. They were bat-like creatures with
leathery wings, and probably exterminated the great flying insects.

There were certainly small mammals in England, as appears
from their jaw-bones in the Stonesfield slate, but if a zoologist
from Mars had visited our planet he would have had to be very

far-seeing to guess that their descendants would replace the reptiles. The first bird fossils have been found in German rocks of this age. The birds in question had poor wings, but long bony feathered tails, teeth, and many other signs of their reptilian origin. None has yet been found in Britain.

The Jurassic strata are exposed over a belt of England lying west and north of the chalk, stretching from north Yorkshire to Dorset. The clay gives flat plains, and the limestones a series of low hills whose steeper faces, or scarps, look on the whole north-westwards. Incidentally they furnish our finest building stone, both from an architectural and a human point of view. Little towns and villages like Burford and Lacock are among the most beautiful in England. And their building did not waste human lives. Masons and quarry-men working in sandstone have a very high death-rate from silicosis, those working in igneous rocks a fairly high one. But limestone workers suffer less than the average from lung disease. In fact one can almost make a geological map of England on the basis of the death-rates of quarrymen in different counties.

Other very important Jurassic formations are the belts of ironstone which are worked at such places as Scunthorpe and Corby. There are various theories as to why a mud containing a lot of iron oxide was laid down at one time and not another. For example there may have been volcanic activity in neigh-bouring regions which brought iron compounds to the surface, and bacteria which concentrated iron oxide as some do in ditches and ponds at present. But no one knows.

There is reason to think that at this time the continents were beginning to drift, and that in particular Europe was moving north from the equator, and South America splitting from Africa to make the beginning of a deep Atlantic ocean. But in Britain at least there were no great changes, even in the first part of the Cretaceous period, when most of Britain was land, though the gault and greensand were being laid down in Kent, Sussex, and Surrey, and other clays and sandstones in Lincolnshire.

This quiet period was brought to an end by a rise of sea level over large areas of the world which is called the Cenomanian transgression, and which drowned England for some forty million years while the chalk was formed.

4

The Chalk and After

ENGLAND, with much of Europe, was under moderately deep sea for about forty million years while the chalk was being formed. It is possible that even Wales was submerged. The chalk contains very little mud or sand from neighbouring lands. But there is more in Devon than in Kent, and there are shore deposits near Exeter, so probably some land remained in western Britain.

There was very little folding or tilting, for the different layers of chalk succeed one another pretty regularly, even though they have been tilted in later ages. However the Chalk Age was most important biologically. One after another the great groups of reptiles died out, often after an apparent outburst of vigour. For example one of the last pterodactyls had a wing span of twenty feet, and was the largest flying thing on our planet before the aeroplane.

The whale-like ichthyosaurs died out in the sea, but other reptilian groups took their places and died out in their turn. The ammonites also came to an end. And on land the dinosaurs disappeared. No one knows why this happened. The Cenomanian transgression might have drowned many dinosaurs, but it could not have drowned the sea reptiles. Changes of climate probably occurred. The formation of the chalk may have removed carbon dioxide from the atmosphere. The evolution of disease germs may have overtaken that of reptiles.

A few biologists believe in a theory of "racial senility", which is like the theory of inevitable cycles of civilisation popularized by Spengler. On this theory it is not clear why lizards, snakes, tortoises and crocodiles are still alive. Whatever the cause, the stage was cleared for the mammals and birds, on land, sea, and air, and in the ensuing period they took the opportunity given them.

It is very likely that during the Cretaceous period North America was drifting away from Europe, and the North Atlantic

Ocean was forming; but certainly if this drift occurred, it went on long after the chalk had been laid down.

The long quiet period of the Chalk Age came to an end with a great folding movement which formed the Alps. Many geologists think this was due to Africa drifting northwards, pushing Europe in front of it and crumpling it. At any rate the Alps were formed by a folding process which has been followed in great detail. The same process produced a series of folds north and west of the Alps, including some in southern England. In the Weald of Kent and Sussex and in the Isle of Wight the chalk was pushed up. In the London area it was folded down.

North-westwards all the strata were tilted, so as to slope down towards London, and ever since then the face of England has been flattened. The hills have been worn down to expose older and older rocks as we go west and north, and a little new land has been laid down in the south-east.

But there has been no great submergence under the sea, and no great folding movements to form mountain ranges. In fact it may one day be possible to give a continuous history of Britain since the end of the Cretaceous period, seventy million years ago.

But just because the history of England in this time has been one of erosion rather than rock or soil formation, it is rather poor in fossils showing how living things evolved during the last seventy million years, comprising the Tertiary period.

The most important beds laid down during this time are the Eocone beds in the lower Thames valley, and in the Hampshire basin, which is the partially drowned valley of a river which ran down the Solent. These clay and sand beds contain a few bones of early mammals. But the main story of how mammals evolved from animals rather like modern shrews into the great variety of living types has been worked out in other countries, particularly the United States.

In the Miocene period, thirty million or so years ago, there was volcanic activity in western Scotland and northern Ireland. Great cracks opened, from which lava poured out over large areas, including most of Antrim and Mull. This may have been due to tension in the earth's crust caused by the drifting apart of

Europe and North America. Fissure eruptions of this kind still occur in Iceland from time to time.

In the succeeding Pliocene period a good deal of clay, sand and gravel was laid down in East Anglia. Its most remarkable feature is a band of gravel and mud several miles across, and best studied near Chillesford, which runs through Suffolk and Norfolk. This seems to be the bed of a great river, which included the Rhine, Meuse, Scheldt and Thames, and emptied into the North Sea somewhere near Cromer.

From this time on we begin to work out the history of our rivers. It is a curious fact that rivers are often older than hills. Our various chalk ranges stand up because chalk is fairly hard, but permeable to water. So it remains when sand and clay are washed away.

Even sand is less eroded than clay, because the water sinks through it, instead of wearing away its surface. For this reason sand often forms low hills near London, such as Hampstead Heath.

The chalk downs are pierced by a number of narrow valleys, some of which still carry rivers, while others are now dry. The rivers were there before the surface had been lowered on each side of the chalk, leaving it as a range of hills.

Once a river is formed, it can easily cut a gorge even through much harder rock than chalk. But it cannot break through a range of hills which is already there, unless its old course is dammed to form a lake which overflows through a pass.

The history of our rivers has been most fully worked out in the chalk country of Kent, Surrey, and Sussex. A number of rivers used to run outwards from a centre near Ashdown forest. For example the Wandle, which flows through south London, used to cross the chalk by valleys at Merstham and Caterham. But the Medway and the Mole were more successful than the Wandle in cutting through the chalk, so they stole the upper tributaries of the Wandle.

In consequence no rivers now cross the chalk at Merstham or Caterham. The Mole is still busy deepening its valley through the chalk. Much of it (as its name suggests) goes underground, and the roofs of these caves will doubtless fall in time. When this happens it may steal some of the tributaries of the Medway, Wey and Arun.

We cannot understand the development of a landscape unless we realise that it is the result of a struggle, not only between the processes which build new land and those which carry it down towards the sea, but even between the different rivers.

5

The Battle of the Rivers

THE history of the English landscape in the last fifty million years is largely the history of its rivers. For there has been no volcanic action and little folding to make new hills, while water has been at work all the time, and ice for a little of it, in wearing the country down.

On the whole the west coast has sunk. There is plenty of evidence that what are now arms of the sea in western Scotland, Ireland, and Cornwall, were once river valleys. So the drainage of England in early tertiary times was more from north-west to south-east than nowadays.

There is a good deal of support for this theory. Thus if we walk along the Cotswolds, we find dry gaps at the heads of the Evenlode, Windrush, and Colne valleys. These rivers run into the Thames from the north-west, and nowadays rise in the Cotswolds. But it looks as if they had once been much longer, but the Severn has captured their head waters, leaving the valleys through the Cotswolds dry. If so there is little doubt that the upper Severn once drained into the Thames by the present Evenlode valley. In fact the Thames rose in Wales.

Many geologists regard this as probable. Buckman went much further, and constructed a map of the ancient English river system according to which even the Ribble in Lancashire and the upper Trent once flowed into the Thames. It is fairly certain that the rivers which run south-west and north-east, like the lower Severn and the lower Trent, are on the whole younger than those which run south-east like the Thames and the upper Trent. The reason is as follows.

In the Midlands there are belts of relatively hard rock which run north-east and south-west, as different layers of chalk, limestone, and sandstone come to the surface. In between them are clays and gravels which have been worn down. The country is like an old piece of timber which was once flat, but where the softer wood has been rubbed away. The older rivers run along the original slope of the flat surface, the newer ones along the softer beds. For example the lower Severn and the Warwickshire Avon run through soft soils, and almost meet the Welland and the Soar, a tributary of the Trent, in the centre of England. The Avon flows south-west, the lower Trent and Welland north-east.

The same kind of thing has happened on a smaller scale in Wales and in Yorkshire. The Swale, Ure, and other rivers used to flow eastward into the North Sea over what are now the chalk wolds. As the softer clays west of the wolds were gradually eaten away to leave the Vale of York, they were captured by the Humber, which had made a deeper valley through the chalk hills.

The Ice Ages, of which the last only ended some ten thousand years ago, complicated matters still further. The ice came from Scotland, the Pennines, and Wales, and on several occasions ice from Scandinavia pushed on to England across the North Sea. This meant that rivers flowing northward were dammed up, and the water had to find its way southwards. In fact there was an occasional reversal of the tendencies which I described earlier.

The most striking case is probably that of the Severn. A glacier from the Lake District crossed part of the Irish Sea and dammed up the Dee and the upper Severn, which then drained northwards. A lake was formed round Shrewsbury, which has been called Lake Lapworth after a well-known English geologist. Its boundaries can be traced by layers of silt laid down in the icy water. It overflowed southwards, and in so doing cut the gorge at Ironbridge through which the Severn still flows. Enough mud, gravel, and boulders were left behind when the glacier melted to prevent the Severn from flowing northwards.

It is still uncertain when the first men came to England. Reid Moir believed that he had found flints shaped by human hands in deposits laid down well before the first ice age in East Anglia. Others think these flints were chipped by wave action. There is however no doubt that men lived in England during the warm intervals between the Ice Ages, including one period much warmer than the present, when hippopotami swam in the Thames. However these men made very crude tools, and were a good deal less human than any existing race. It was not till shortly before the last of the four Ice Ages that the modern type of man appeared in Europe, with a greatly-improved flint-chipping technique, and, what is perhaps more striking, with the habit of making sculpture and painting. Some of them got at least as far north as Derbyshire before the last advance of the ice drove them south again.

During the Ice Ages so much frozen water was piled up on the land that the sea level was much lower than now. England was connected with France, and the Thames flowed through the site of London in a channel now buried under gravel and mud below sea level. The straits of Dover were only finally formed when the sea level rose again as the last ice sheet melted. England was colder than now, but many men had already come there from Europe, and others followed in small ships.

This very brief sketch of our country's history should not satisfy most of its readers. Every educated person should know at least something of the geology of his or her own district. The Londoner should know that the clay in his garden was laid down under subtropical conditions, so that it may contain relics of banana trees, crocodiles, and tapirs. The citizen of Edinburgh can be equally thankful that the Castle Rock is no longer a volcano, and that the ice is no longer cutting its western face, and piling up the boulder clay of the Royal Mile to its east. The Welshman may be proud that three great systems of rocks, the Cambrian, Ordovician, and Silurian, were first identified in his country and named after its inhabitants.

All of us can learn something of the history of our country, and see to it that at last men can begin to shape it to their own use instead of leaving it to the blind play of geological and economic forces.

II
GEOLOGICAL ESSAYS

I

Dating the Past

UNTIL a generation ago, the only events which we could date accurately were those which were part of a history where each year had a numerical date such as the year 412 after the founding of the city of Rome, or a date by means of officials, such as the year when Caesar and Bibulus were consuls. Even where the lengths of kings' reigns were accurately recorded, one does not know if "ten years" means ten years and one day or ten years and three hundred days, so the uncertainty soon piles up. About thirty years ago, a number of historic eclipses were accurately dated, so that we know the date of the siege of Troy within ten years, and, what is more remarkable, that the city of Ur in Iraq was destroyed by the Elamites in B.C. 2283. This is still the earliest dated event in human history.

Where there is no written history, there are two methods by which we can give an exact date to buildings or tools, which tell us a good deal more about men than lists of kings and battles. One method is by tree rings. In a country like Arizona, where there are so many dry years that the trees are growing under a severe handicap, far more wood is formed in a wet than a dry year. By examining recently cut trees one gets a calendar for several hundred years back.

Douglass, the pioneer in this investigation, was not content with this. He examined tree trunks from abandoned "Indian" buildings which had been preserved in the dry climate. He found some whose ring pattern overlapped that of living trees. That is to say their outer layers showed the same succession of broad and narrow rings as the insides of trees recently felled. From these he was able to work still further back in the same way until the earliest date determinable is about A.D. 400. So he was able not only to date houses and tools, thus making a cultural history possible, but other workers, by examining giant Sequoia

31

trees, have obtained at least a rough history of the weather of California for the last 3,000 years.

In wetter countries, annual layers of mud are laid down in some lakes. When heavy rain or a big thaw brings down a lot of mud, the layer is thick. In this way different mud sections can be compared and finally synchronized. And where human tools are embedded in the mud they can be dated. This method of dating by annual mud layers works best in the neighbourhood of retreating ice sheets. It has been particularly used by de Geer to date mud layers or varves, as they are called in Sweden. But unforunately men were rare or absent in the neighbourhood of these giant glaciers, so such records can only be used rather indirectly to date human events. Still they do give us the earliest dated event. In the year 7912 B.C. the freshwater lake occupying the northern Baltic basin, but considerably above the present sea level, burst across southern Sweden in Vestergötland, opening a valley which lowered the lake level by 90 feet. The present connexion further south was only opened later, when the sea level had been raised by melting ice.

We have a pretty good record of successive events during the ice ages. For example it is quite clear that there were four main cold periods with warm ones between them. But whether the last one went on for fifty or a hundred thousand years is not so certain. It may be certain a generation hence. The Yugoslav scientist Milankovitch claims to have dated the ice ages on the basis of an astronomical theory. At present most geologists do not accept his calculations. This does not mean that they think he is wrong, merely that they are not convinced. In another twenty years they may be convinced. If Milankovitch is right, we can date the big climatic changes of the last half-million years with errors which vary between five per cent. and twenty per cent. The ice ages began about 600,000 years ago according to his calculations. Other estimates give over 700,000 years.

The method of annual mud layers can be used to date sections of the past. Sometimes the evidence for a yearly cycle is overwhelming. For example fossils of adult insects are found in one part of each layer, and of their larvae in another. Bradley counted enough varves in the Eocene formations of Colorado, Utah, and Wyoming, to be able to say that a particular epoch lasted between five and eight million years. During this time a thickness of about two thousand feet of sandstone and oil shales were laid down in two lake beds. But this only gives us the duration of about a third of the Eocene period, or a tenth of all the time which has gone by since these rocks began to form. Other estimates are based on the rate of animal evolution.

Far the best dates for remote events are given by radioactivity. If a rock contains uranium, just one atom in 6,578 million of this element is transformed every year. Most of them pass through the stage of being radium, a few through that of being actinium, and in each case end up as lead. A very tiny fraction split in two, as in atomic bombs. Thorium, another radioactive element, behaves in a very similar way. Thus, if a mineral was formed containing uranium, but no thorium or lead, one could calculate its age from the amount of lead contained in it and produced since its formation. However, this would be almost useless in practice, because one could never be sure that some lead had not been present in the original mineral. Fortunately however the lead derived from radioactive atoms has a different atomic weight from ordinary lead. So it is possible to find out how much of the lead in a mineral was derived from the uranium or thorium in it, and thus estimate its age exactly.

On this basis enough rocks have been dated to fix the dates of all rocks carrying fossils with an error which rarely reaches ten per cent. We thus have at last the time scale of evolution. We know that about 500 million years have gone by since the earliest rocks were formed containing well preserved animal remains in any quantity, and 400 million since the first record of vertebrates. 270 million years ago our ancestors left the water, and 70 million years ago the mammals took over from the giant reptiles and became the dominant land animals.

The complete story of how these figures were reached is of course, a very long, but extremely interesting one. Any good public library should contain Zeuner's *Dating the Past** and Holmes' *The Age of the Earth*†; and Zeuner's series of clues make

* Methuen 1946.   † Nelson 1937.

C

even the best detective stories look pretty elementary. Most of his ideas are generally accepted; however Milankovitch's theory is still on trial. And in 1946 Holmes, as a result of more lead analyses, has arrived at three thousand million years for the age of the earth, whereas the oldest known rocks are little more than half this age.

Later I am going to try to tell something of what has happened in this vast period of time, in order to help my readers to form a picture of our world in time, like the picture which they form in space when they learn geography.

2

Geological Revolutions

THE first geologists thought in terms of the accounts of the earth's origin given in the Bible. The world had been made, a few thousand years ago, in seven days. Since then there had been very minor changes due to erosion by water, to volcanoes, earthquakes, and so on, with the one exception of Noah's flood which had covered the whole world. When great depths of mud and rock were found, containing bones and shells of extinct animals, these were supposed to have been caused by this one flood. So almost the whole of the geological record was explained by one single catastrophe.

More careful examination showed that many rocks must have been laid down very slowly in still water, and that there had not been one set, but hundreds of sets, of extinct animals. In fact a hundred years ago geologists generally thought that the rocks had been gradually laid down in an orderly way. This theory works pretty well for most of England. If we find one deposit above another, for example the London clay above the chalk, we can be sure enough that the lower one was formed first. It does not work for the great whin sill in Northumberland, which is due to an underground flow of lava which has baked the sedimentary rocks above and below it. Of course a surface flow only bakes the rocks below it, as there is nothing above to bake.

But where great mountain ranges are investigated we find old rocks pushed over newer ones, and sometimes whole strata turned upside down. This was first detected in the Alps, but later in Scotland and Belgium where the remnants of mountain ranges much older than the Alps, and almost completely worn away, were discovered. We now know that it is pretty general.

Modern geologists think in terms of revolutions. The word is not mine, but theirs. There have been about three of these revolutions in the last five hundred million years, and about eight in the earlier history of the earth. Each revolution built a number of mountain chains in different parts of the world. The last one produced the Himalayas in Asia, the Alps in Europe, and the Cordilleras in America. Some geologists would divide it into two stages. A little mountain-building is still going on, but we seem to be past its climax, and we may expect that during the next hundred million years or so mountains will be worn down rather than built up.

The most violent revolutionary changes are to be looked for in the building of the highest mountain chain, the Himalayas, and Dr. E. B. Bailey, the Director of the British Geological Survey, has recently interpreted the findings in this range of two Swiss geologists, Heim and Gansser, and the Indian geologist, Wadia. Little is known of the geology of the Himalayas, since the Tibetan and Nepalese governments do not want their countries to be developed by capitalist imperialism, as they might well be if valuable minerals were found in them. The most interesting area which has been fully studied is that around Kiogar, between the upper courses of the Ganges and Sutlej. Where valleys are several miles deep the opportunities for working out the structure of the mountains are of course very good.

It becomes fairly clear that Tibetan mesozoic rocks with fossils like some found in Bavaria were pushed bodily over younger rocks formed a long way further south. In between them there are volcanic rocks, but telling a very queer story. The lower layers consist of lava with the characteristics only found when it has flowed out under the sea, interspersed with the skeletons of one-celled animals which only accumulate in deep oceans. But above these are volcanic rocks of a terrestrial type, pushing up through the sediments above them in narrow veins like the quartz veins which are so common in Cornwall and many parts of Scotland. In fact Tibet has been pushed over India, which has slipped below it and lifted it up. In the early stages there was a deep and narrow sea in front of the advancing edge of Tibet, studded with underwater volcanoes, such as are not rare in the Pacific. Later these were overrun. From a geological point of view the Himalayas at Kiogar are in an almost undescribable mess. Huge blocks of rock several miles long have been torn away, perhaps by landslips, and embedded in younger strata.

Dr. Bailey has not yet unreservedly accepted the theory which is to-day the great geological heresy, but may be generally accepted in another generation, the theory that the continents have drifted for thousands of miles over melted matter below them like slag on the top of a mass of molten metal. Its strongest adherent to-day is probably the South African geologist du Toit.

On this theory Africa, South America, India, Australia, and Antarctica were once joined together. They have drifted apart, and the pressure of Africa against Europe made the Alps and Pyrenees, while the northern tip of India went bodily under Tibet. This theory certainly explains a lot of facts. For example the world's main deposits of diamonds are in South Africa, and in the part of Brazil which once fitted onto it. The ice-formed rocks in India, Africa and Australia which are now in hot regions were laid down by glaciers near the South Pole, and so on.

Geologists are becoming more and more ready to believe in very large changes in the earth's surface in the remote past. Much the same changes have taken place in anthropological theory. We no longer ascribe all our evils to a woman eating an apple in a garden six thousand years ago. But we can attribute many of them to the origin of class society first in one part of the earth and then another, during the last ten thousand years. This meant division of labour and greater production, but it also meant a huge growth of oppression and selfishness. We can no more accept the details of the garden of Eden than of Noah's flood; but we can agree that there have been rapid and world-wide changes in geography, and, at a very much later date, in human society.

3

Earthquakes

THERE has been a terrific earthquake in Ecuador, and most of the inhabitants of several cities have been killed. There are several questions to be asked about such an event. Why did it happen? Could it have been prevented? If not could it have been predicted, and measures taken to minimize the loss of life? Is anything going to be done to prevent similar disasters?

Earthquakes have a strongly marked geographical distribution. Most of them occur round the shores of the Pacific Ocean, though there are other regions where they are fairly common, notably parts of the Mediterranean basin, and the north Indian plain. Along most of the Pacific shores mountain chains have recently been formed, and there is good evidence that they are still being formed, by a process of folding. Corresponding to the mountain ranges parallel to the coasts there are very deep troughs in the sea quite near the coast.

Whereas on the shores of the Atlantic the mountain ranges seldom run parallel to the coast. And except off the West Indies the sea is seldom very deep near the land. The Pacific coasts are under pressure. Apparently the continents are being pushed towards the ocean, and the folding is a consequence of this. In Ecuador, and in some other regions of folding, there are volcanoes.

But it seems probable that the earthquakes cause the volcanoes, and not the other way round. The folding cracks the rocks so that the molten materials underneath seep up at points of weakness. Volcanoes also form at points where the earth's surface appears to be under tension, and tearing rather than folding. Volcanic regions of this sort are characterized by fissure eruptions when a relatively broad crack opens, and lava floods many square miles. Iceland is a region of this kind.

Earthquakes also occur in regions where new soil is being very rapidly laid down. These may be at the foot of great mountain ranges like the Himalayas. Here the rivers are laying down silt so quickly that the rocks below them have not had time to bend under the extra strain, and may crack. The newly formed soil is far from stable, and fissures open in it from which mud and sand pour out. Earthquakes occur for a similar reason a long way from mountains along the coast of the Gulf of Mexico, where the Mississippi and other rivers are laying down great quantities of mud on the shores. But this type of earthquake is generally less serious than that associated with folding and mountain building.

After an earthquake on the Pacific shores one generally finds that there has been movement at a fault. This is a crack where the even sequence of the stratified rocks is disturbed. After an earthquake a survey will show that the ground on one side has moved perhaps six feet northward and two feet upward relative to that on the other side. Faults are quite common in coal measures. A coal seam suddenly comes to an end in rock, and may be found again many feet above or below the old level. Such faults are almost certainly relics of earthquakes many millions of years ago, which built the hills of Britain, or at least those of them formed since the coal was laid down.

In many areas round the Pacific, but especially in California, the faults formed by earthquakes in the last century are quite well known; and since there have been repeated movements of some of them, it is obviously dangerous to build near them. So there is a constant struggle between geologists and real estate speculators who want to get houses built in their neighbourhood and sell out before the next earthquake. On the whole the real estate men win out. So a good many houses around Los Angeles, for example, are not likely to last out the next century.

Not only can one avoid the neighbourhood of faults in buildings, but one can build earthquake-resistant buildings. They should be steel-framed, and not too firmly anchored to the ground. Unfortunately many buildings in Ecuador, and especially the churches, are modelled on those of Spain, where earthquakes are rather rare, and are deathtraps in a less stable country.

Earthquakes seem to be a little commoner when the moon

is full or new, and the extra pull of the moon, which causes spring tides, is added to the strains already existing in the earth. But one cannot yet predict just when an earthquake will occur in a given place. The most hopeful apparatus for prediction, though not a very hopeful one, is an aquarium. Some Japanese biologists have stated that certain fish show signs of unease for an hour or two before an earthquake. This is not impossible. Fish have special organs for perceiving vibrations in water, which help them to detect their prey or to escape from enemies. They may notice very faint tremors preceding a great earthquake. If they do, an apparatus could probably be made to do it better. But I fear that either fish or any machine which we could devise would give a good many false alarms.

The one efficient precaution against earthquakes is to make buildings as near as possible earthquake-proof, and to build cities in unstable regions so that fires will not spread; for water-mains are pretty sure to be broken by a severe quake. Very few people are killed in the open except by landslides, floods, or sea waves, and these dangers could largely be avoided by siting houses properly. Unfortunately all these precautions involve planning, and interference with private enterprise.

There may be more important things to do in Ecuador than make earthquake-proof buildings. Probably the total number of people killed there by earthquakes in the last century is quite a small fraction of the number killed by easily preventable diseases. A hundred people killed by an earthquake at least get a mention in the newspapers. A hundred babies dying of diarrhoea whose germs are carried by flies do not. House-flies kill more people than earthquakes, and mosquitoes a great many more. So probably the people of Ecuador would do better to spend their public funds on proper drains and water supply than on earthquake-proof buildings. Unfortunately their country is so infested with generals that they are more likely to spend it on armaments than on either of these necessities.

4

Oil under England

NOW that the danger of bombing is over, the location of the British oilfields which were developed during the war has been revealed in an article in *Nature*. The main one is in Nottinghamshire, at Eakring and other villages, about eight miles north-west of Newark, and there is a smaller one at Formby in Lancashire, between Liverpool and Southport. Besides this, natural gas has been found near Dalkeith, a few miles south-east of Edinburgh, and at Aislaby in north Yorkshire. The oilfields had yielded 337,000 tons, or about ninety million gallons, up to the end of 1944. The gas has not yet been used. The oil was discovered by the D'Arcy Exploration Company, a subsidiary of the Anglo-Iranian Oil Company. The Newark oil is about 2,000 feet below the surface, the Formby oil at a depth of only 120 feet.

The Newark oilfield, like most others, is situated in what is called an anticline, that is to say an area where strata, which originally lay flat, have bulged upwards. If a layer of porous sandstone contains oil, but a layer of less permeable rock lies above it, the oil will rise as high as it can in the sandstone, and there form an underground pool or swamp. If a borehole is put down through the hard rock above it, the oil may rise to the surface, or, as in the English fields, pumping may be necessary. The most striking anticline in England is the Weald of Kent and Sussex. This has yielded no oil, and only enough gas to light the railway station at Heathfield for a few years. The oil-bearing anticlines were buried under other deposits which lay comparatively flat on top of them. The Lancashire oil sandstones were under clay laid down during the Ice Age. The Nottinghamshire field was covered by a variety of rocks laid down, mostly under the sea, since the sandstones were folded. We know the surface structure of England and Scotland well enough. But only in some areas do we know anything about the depths.

Here is the reason. Sometimes many beds of mud or sand were laid down in the same area, so that they lay evenly on top of one another. Then even if they were later folded, tilted, and partly worn away, one can tell how they are arranged below the surface by studying those which are exposed, just as one can tell how the grain of a plank is arranged in its interior by studying its surface. But now think what would happen if the sea rose and covered much of Britain, as it will do if the ice covering Greenland and Antarctica melts. Silt would be laid down, gradually filling in the valleys to form a nearly level plain. If this were raised up again above sea level it would be impossible, from a study of the surface, to say whether a boring under a given place would reach chalk, sandstone, clay, or coal measures. A break of this kind in the geological record is called an unconformity.

The oil-bearing sandstones date from the time of the coal measures or a little later. There was a big folding of the rocks and building of mountains just after the coal was formed, and in many areas the coal measures were washed away before any later rocks were laid down. So there is coal under parts of Kent and Southend, but not under London. The oilfields were largely discovered by letting off explosives at one point, and measuring the time taken by the shock wave to reach other points. Since the wave travels quicker in dense rocks than in light ones, it is refracted back to the surface after going down below ground, and this gives rough information of the depth at which denser rocks are found. Borings were then made on the basis of the results of these artificial earthquakes.

Besides oil and gas, a coalfield was discovered near Lincoln, and potash beds in Eskdale, Yorkshire, both at about 4,000 feet, as well as a deeper coalfield near Nottingham; and a large amount of geological information of no immediate economic importance was obtained.

It may be that Mr. Lees, of the Anglo-Iranian Oil Company, who was in charge of the work, will publish the results in the fullest detail. But the country needs a complete survey of this kind. The Reid report calls for an exploration of our coal resources. We ought to know about all the mineral deposits within a mile of the surface. No investigation carried out with reference to one particular mineral, whether it be oil, coal, potash, tin, or copper, will give all the needed information; and it is very undesirable that any particular company or trust should have the first call on knowledge of our national wealth, let alone possess secret knowledge. The organization which should undertake work of this kind is the Geological Survey. It could not have done so in the past because it has been starved of money, personnel, and equipment. Now that mining royalties have been nationalized, it is essential that we should know what our mineral resources are. This will be even more urgent when the mines are nationalized.

The experience of the Soviet Union shows that nationalization is not enough without full scientific knowledge. We need a big programme of geological research with artificial earthquakes and deep boreholes all over the country. There should now be plenty of explosives to spare for the former purpose. Similar surveys are needed in the colonial empire. Unless our surveys are expanded rapidly we shall have dumped our spare explosives in the sea instead of using them for research. It pays capitalists better in peacetime to prospect in countries where the miners are paid very low wages. It is therefore important that the miners' M.P.'s in Britain should press for a big extension of geological research in our own country. They will be told that it is a waste of money. But when we needed oil during the war, it was found. We need a scientific exploitation of our underground wealth if we are to win the peace.

5

Exploring Subterranean England

NO one can understand the social conditions in England to-day, unless they know about its past. That is why we ought to learn history. Similarly we cannot understand its surface features unless we know about the rocks beneath.

Sometimes a study of the surface rocks tells us a good deal about what is below them. For example we find the chalk beds dipping down gently in the Chilterns to the north, and the North Downs to the south of the Lower Thames. And it is reasonable to guess that there are chalk beds under London joining the two areas where the chalk comes to the surface. Actually a lot of borings show that this is true. Where the hidden beds contain something more important than chalk, such as coal, this kind of evidence is generally valuable. But sometimes it lets one down. For example the coal seams in the Scots Lowlands have often been destroyed by volcanoes such as those of which Edinburgh Castle and Arthur's seat are the cores. But borings are very expensive. There are several ways of finding out what is under our feet. One is by small artificial earthquakes. This method is used by the oil trusts, but they refuse to tell the scientific public what they have discovered, or even how they discovered it. So one of the departments of Cambridge University has had to break a hole in this "Iron Curtain".

The first results of their work were published by Bullard, Gaskell, Harland, and Kerr-Grant during the war, and dealt with investigations made round Cambridge. In case I am accused of bias, I quote from their introduction. "Unfortunately, most of the work has been carried out by commercial firms which do not publish detailed accounts of their methods or results. Anyone wishing to use the method has therefore to start practically from the beginning, and solve problems which have already been solved several times before."

Here is the problem which they studied.   Under the chalk in eastern England are clays such as the Oxford clay, and sands, all laid down in shallow water with thin beds of limestone.   Below these is a great variety of much older and harder rocks, laid down in palaeozoic times, and including coal measures.   When England next sinks under the sea, as it probably will in the next hundred million years, mud will be laid down on the submerged surface, and the valleys gradually filled up.   The new beds will sometimes rest on recent deposits like the Thames gravels, sometimes on far older ones like the old red sandstone of Devon. This kind of surface, where newer rocks rest on a variety of old ones, is called an unconformity.   Bullard and his colleagues mapped the unconformity below the clay and chalk, that is to say the surface of the Old England which sank under the sea nearly two hundred million years ago.

This is how they did it.   They had six geophones buried in the ground.   A shock in the ground drives an armature nearer to a magnet and induces a current in a coil wound round it.   This is recorded on a moving photographic paper by an oscillograph, along with the record of the break of the current in a wire wrapped round a charge of gelignite.   From a quarter of a pound to fifteen pounds of gelignite were buried at distances of a thousand to eight thousand feet from the recording instruments, and the times taken by the shock waves were read from the paper.   The biggest source of trouble was cows, which chewed the wires.

Now a shock wave travels at only about 1,000 feet per second in surface sand, 7,000 feet per second in chalk, and up to nearly 20,000 feet per second in very hard rock.   So if the charge and recorder are far enough apart, the shock will cover the distance between them in less time by going down to the hard rock than by travelling below the surface.   From six or eight records at different distances one can determine the depths at which hard rock surfaces are reached, and even the slopes of these surfaces. Of course the first thing to do was to see whether the depths so found agreed with those actually found in boreholes.   Fortunately they did so quite well.

The results can therefore be trusted.   They show for example that the hard rock floor on which the clay, sand, and chalk were laid down, and which comes to the surface in western and northern England, is less than 400 feet below sea level at Cambridge, compared with 1,000 feet or more at London and at Corby. In fact contour lines have been drawn, and we have a rough idea of the hills and valleys in this part of England some two hundred million years ago.

More remarkable, we know something about the actual nature of the rocks from the speed at which shock waves move through them. Below Corby the waves move very quickly and the few borings show volcanic rocks. Further south and east the waves travel more slowly, showing softer rocks of at least two different sorts. The area covered is not one where coal is at all likely to be found, nor is it likely to reveal other valuable minerals. However we can say that if there are non-ferrous metals anywhere in this area, they are more likely to be under Corby than under Cambridge or London. But similar work could be done to determine the probable boundaries of coalfields such as that of Kent, where the coal seams never come to the surface. It could also be done under the sea, where we already know that coal seams run out for many miles.

Such work, if done by "free enterprise", is not published. If it is to be of value to Britain it must be done either by a state organization, such as the Geological Survey, or by universities which, in practice, rely on state subsidies. The time has gone past when a geologist needed no tools but a hammer and surveying instruments. All Bullard's apparatus could be carried in a single van, and doubtless cost less than a single aeroplane. But they could not be made in an ordinary workshop.

His work was probably of no immediate economic value, though it may pay its way many times over if only by suggesting where not to bore for coal, oil, or iron ore. Similar work elsewhere may be of great value. Admittedly our immediate planning policy must be a short term one to get us out of our present economic difficulties. But we shall need a long term policy, and this is only possible if we know our mineral resources. Here is one way to discover them.

6

I Ask for Fossils

IN the last hundred years it has become constantly harder for ordinary workers to contribute to scientific research. The great physicist Faraday started off as a bookbinder's apprentice and went straight into research work. A bookbinder's apprentice wouldn't find it so easy to-day. But there is one important contribution to knowledge which can only be made by coal miners. That is the collection of fossil animals from the coal measures. This article is a call to miners to help in the work. Those who do so can not only contribute to science, they can earn good money.

Everyone knows that the coal seams consist of the remains of vegetation which grew in swamps. Very few bones or shells are preserved in the coal itself, because rotting vegetation produced acids which ate them away. But the newly-formed coal was often submerged under sand or mud, which hardened to sandstone or shale, to form the roof of the coal seam. In these rocks fossils are sometimes found.

Now shells, resembling those of modern mussels, cockles, scallops and snails, are fairly common in the coal measures. They have been used for dating the rocks. While they are quite interesting, I doubt if anyone will buy them. On the other hand the remains of vertebrate animals are of very great interest and some monetary value. By vertebrates are meant animals with backbones, including fish, amphibians, reptiles, birds and mammals. Any bones or teeth, and most scales, come from vertebrates.

So far only two classes of vertebrates have been found in the coal measures, namely fish and amphibians. The fish were not so very different from modern fish. The amphibians were mostly four-footed animals, rather like modern newts. But some of them were much larger. The smallest were only about three inches long, the largest up to twelve feet, as big as modern crocodiles. Occasionally a complete skeleton is found, more often a skull or a few vertebrae from a backbone, most usually only single scales or fragments of bone. The fish are far commoner than the amphibians. Both these classes of animal are very interesting to the student of evolution for this reason. The ancestors of the amphibians were fish which developed legs in the Devonian period, before the coal measures were laid down. But the amphibians in the coal measures were much more like fish than are modern frogs or newts. And some of the fish of the coal measures were much more like those which gave rise to the amphibians than any modern fish. So by studying them we can learn something about the change by which the first four-legged animals arose.

"Is it not possible," a reader may ask, "that one might find the bones of some much more highly developed animal such as a bird, or something like a horse or a dog, in the coal measures?" The answer is that it is not impossible. But if such a fossil were found, it would disprove the theory of evolution. For biologists believe that they have found the fossil bones and teeth of a great many of the steps by which mammals and birds evolved from reptiles. And these are found in rocks laid down long after the coal seams. So such a discovery would be most surprising.

In spite of their interest, we know very little about British vertebrates from the coal measures. A few have been found in Scotland. Ninety per cent. of the English ones come either from Newsham pit in Northumberland, which is now closed, or from the southern end of the North Staffordshire coalfield. The man who was mainly responsible for the collection at Newsham was T. Atthey, a grocer. He bought them from miners in return for credit at his shop, and sold them to museums and to palaeontologists, probably making a very good profit out of them. J. Ward, in Staffordshire, certainly collected himself, but also seems to have bought specimens from miners. Both these men lived in the nineteenth century, and collected fossils over a period of about thirty years. During the present century nothing has been found but a few fish-bones.

The reason for the falling off is perhaps that nowadays there are fewer amateur fossil collectors than eighty years ago. The collection, and still more the preparation, of fossils is a highly skilled business. The amateur is less likely than he was to find anything fresh in most places. But this is not true of coal-mines. To-day the coal miners are far better educated than their grandfathers, and they are just as likely to find something of real scientific interest.

Where should they look? The vast majority of vertebrate fossils occur in cannels, in cannel-like shales, or in nodules of ironstone imbedded in thick shales. They are most likely to be found in the roof of a coal seam, but they may be found elsewhere. So far as we know they may occur in any coalfield. The southern end of the North Staffordshire field is perhaps the most hopeful area. But fossils found there are likely to be of kinds already known, while those from other places will more probably be of hitherto unknown animals.

Unfortunately the fossils occur in small patches, perhaps where a pool containing a number of fish dried up, and their skeletons were buried under mud. Such an area is only worked for a short time. It is very unlikely that any particular reader of *Coal* will find any. But if ten thousand miners start keeping their eyes open, it is pretty sure that a dozen of them will find some fossils.

In order to start the ball rolling, I will offer £5 for the first fossil fish or part of a fish sent to me, and £10 for the first fossil amphibian or part of an amphibian.* If the find is worth more, that is to say if the British Museum or some other museum will give more for it, I will hand over the price, and the same with any later finds. It is essential that the place of finding should be accurately stated (e.g. roof of the —— coal seam, about —— yards N.N.W. from the —— shaft of the —— pit. If the specimen is found among shale in a dump or on the surface, the name of the pit should be given, and any possible indication as to where the fossil probably came from. My address is :

Prof. J. B. S. Haldane, F.R.S.,
Dept. of Biometry,
University College,
Gower Street, London, W.C.1.

* The £5 has been paid. Unfortunately the £10 has not.

D

I am not much of a palæontologist myself, but I have some younger colleagues who are experts on fossils, and are out to help me. In fact the scheme is theirs, and not mine. I have also arranged with the editor of *Coal* to publish the names of the first finders, and of anyone who makes a notable find later on, unless they ask that their names should be kept back.

The value of a fossil is anything from about a shilling for a fish scale or tooth, up to several hundred pounds for the complete skeleton of a new type of amphibian. Apart from the financial value and the knowledge that one is helping science, the finder of a new species is usually commemorated in its name. Supposing a miner called Evans finds the skull of a new kind of *Orthosaurus*, an animal rather like a small crocodile, it will probably be called *Orthosaurus evansi*. Or if he preferred to commemorate the name of Will Lawther or Arthur Horner, it could be called *lawtheri* or *horneri*.

What I hope to do is to inaugurate a regular scheme of purchase from miners. Such schemes have worked very successfully elsewhere. Most of the fossils from the English Chalk have been found by quarrymen or lime-kiln workers, who have sold them to scientists. The great Thomas Henry Huxley got quite a number of fossils of amphibians from the coal measures from miners or surface workers who knew of his interests. But unfortunately, there seems to be a gap between miners and palæontologists to-day.

It is quite possible that some miner may strike a rich deposit of fossils, and become a real expert himself. This has happened in one case. David Davies, a Welsh mine foreman, was the first to make really large collections of plant remains from different coal seams. He showed that even where the plants did not differ very much, there were differences in the proportions of different kinds, just as in one meadow you will find a great deal of clover among the grass, in another very little. I am glad to say that the University of Wales gave him an Honorary Degree for his work. Unfortunately, plant remains from coal seams are pretty common, and I am not prepared to pay for them, though I would certainly pay for insect remains, if any were found.

The nationalization of the mines is only a step towards socialism, though it is a big one. Socialism in the full sense will only be possible when the workers in every industry understand enough about the conditions of their work to be able to control it, and in particular when most experts, such as engineers, geologists and medical officers, are drawn from the ranks of the workers. Now the study of fossils is absolutely essential for geology, because the different rocks are dated by the fossils in them. Naturally enough the geologists use the commonest fossils for dating. And the commonest fossils are generally shellfish. But shells are much less interesting to the student of evolution than bones. One cannot tell much about the animal that made it by looking at a shell. One can tell a great deal from bones. Some of the amphibians in the coal measures had lost their legs, and degenerated into eel-like creatures which could not come out of the water. Others had powerful legs and could lift their bellies off the ground. In a few cases from the French coal measures, we have enough specimens of a species to know that they started as something like tadpoles and only developed legs as they grew up. Again the teeth show us what kind of food they ate. And just as some of the shells enable an expert to date a rock very accurately, the vertebrate fossils are characteristic of longer periods. If you show me an oyster shell I have not the least idea of its age. The oyster has not evolved much. It has stayed put for hundreds of millions of years. But if you show me three skeletons I can say that this one is probably a fish from the Old Red Sandstone, that one a reptile from the Jurassic and the other a mammal from the Eocene. So to get a broad view of geology one must study the vertebrate fossils, which show a fairly steady progress, as well as the invertebrate ones. Miners can begin to learn geology by studying the fossils from their own pit if they are lucky enough to find any. And by doing so they will be helping to raise the miners as a whole to the level of knowledge where they can take over the management of the mines completely.

I know that many people think that science should be severely practical, and that the detailed study of fossils is of no practical importance. This is quite untrue. Any bit of "highbrow" science may prove to be of the greatest practical value. For example the British coalfields were laid down near the sea. Some of those in France and all of those in Czechoslovakia were laid down in lakes well away from the sea. Naturally they contain very different fish and shells. So it is hard to say whether a Czech coal seam is earlier or later than an English one. But an insect could fly or be blown from one coal swamp to another. So we might be able to find out which seams were formed at the same time if we found the same insects in both. This sort of dating tells us about the structure of Europe in these ancient times, and suggests where to look for new coalfields.

But I believe there are a great many miners who are interested in knowledge for its own sake. It is just as interesting to know what fishes lived in Dinantian times (the technical name for the group of some million years during which some coal seams were formed) as to know who won the cup final in 1937. Those who laugh at this kind of knowledge are simply trying to prevent their mates from knowing more than themselves, in fact keeping them down. They are playing into the hands of those who don't want the workers to have free access to all kinds of knowledge. By looking for fossils in your coal pit you will not only be helping science and perhaps earning some money. You will be helping to raise the status of your profession, and to break down the division of classes in British society.

7

I get Fossils

LAST month I wrote an article in *Coal* asking coal miners for fossils of certain kinds, and I am just beginning to get the results. Before saying anything about them, I want to say what I asked for, and why.

Fossils are of interest for two rather distinct reasons. In the first place they tell us about animals and plants which lived in the past, what they were like and how they had evolved. Secondly, they enable us to date rocks. Two beds which contain just the same set of fossils must have been laid down at much the same time, for evolution goes on quickly enough to produce marked, though not very striking, changes in half a million years; and the majority of rocks were laid down at a rate of less than 100 feet per million years, often very much less.

In just the same way a student of ancient coins may use them to study the development and degeneration of metallurgy or craftsmanship. Or he may use them for dating. For example in the cave called Wookey Hole, in Somerset, coins have been found which were made under 17 Roman emperors who reigned between A.D. 60 and 392, but nothing later until quite recent times. Clearly people lived there up to about A.D. 400, possibly refugees from the troubles which occurred when the Roman legions withdrew; but it was not inhabited in Saxon times or the middle ages.

Now I am trying to get fossils from the coal for my colleagues Kermack and Kühne, who are palæontologists rather than geologists. That is to say they are interested in fossils for their own sake rather than for dating. In order to date rocks you had better study the commonest fossil species, which are generally molluscs or other similar shellfish. The commoner types of shell from the coal measures are pretty well known. They are economically important because they help to identify corresponding coal seams in different areas. The fossil plants are also fairly well known, partly because there are a great many of them, as is natural since coal consists of plant remains, partly because they throw some light on how coal was formed, and have therefore been studied.

However, what my colleagues want are fish and amphibian bones and teeth, as they tell one a great deal more about the animals to which they belonged than do shells of animals resembling mussels. I have offered to pay for fish or amphibian remains provided I am told just where they were found, so that we can organize a search for more, if anything interesting is found.

Unfortunately my first few parcels have mostly consisted of plant remains, with a few molluscan shells. No bones have yet turned up. There is however one beautiful little animal related to the living king-crab. It looks rather like a very large wood-louse. However, as a matter of fact it was more nearly related to the spiders and scorpions. I shall certainly sell it to a museum and let its finder have the price, though I fear it will not fetch more than a pound or so.

At least one of those who have sent me fossils obviously knows something about the subject, and with a little luck may get something important. I am afraid some of the others will regard me as a swindler because I am not prepared to pay for shells of molluscs, even though they are called shell fish.

I did my best to make it clear just what I wanted, but I obviously did not succeed very completely. Also some of my correspondents may have mistaken tree bark for fish scales. This is not their fault. Palæontology is not taught in schools, and what is worse, men and women who regard themselves as educated are often totally ignorant of it. This is largely because our educational system is totally pre-scientific. Our wretched school children have to learn whom Edward II married, and why this provided Edward III with an excuse for invading France. They have not the vaguest idea what their ancestors looked like fifty million or two hundred million years ago. I don't regard the ancestry of the human race as certainly established, but it is better established than the legitimacy of Edward III, to judge from what has come down to us about the private lives of his parents.

Even if I became Minister of Education, we could not start

teaching children palæontology because it is iquite possible to become a qualified teacher of science without knowing anything about it. But I should see to it that future teachers learned a little of this science, even if they had to miss the Kings of Judah and Israel.

Above all, miners and quarrymen ought to know some, just as engineers ought to know some physics and chemistry, farm workers some biology, and seamen some meteorology. It is not merely that it may be useful. A self-respecting man or woman ought to understand what he or she is doing. Otherwise they are half way to being slaves.

Of course there are good reasons why workers are discouraged from gaining such knowledge. For one thing it would put them on a level with much more highly paid experts. For another they would probably learn about the economics of their job as well. And this kind of knowledge does not make for the stability of capitalism. But I want to see the miners learning all that is necessary to take over their industry completely. Economics is one thing they will have to learn. Palæontology is another.

8

We are still Suffering from the Ice Age

THE thaw of 1947 brought floods to many parts of England, but probably the worst were in the fens, and in the area between York and Doncaster. Elsewhere they were severe, but did not last so long.

The reason is clear enough. The flooded areas are among the flattest parts of England. But the reason why they are so flat is interesting. When a land emerges from under the sea, rivers gradually form in it, and cut valleys which generally have a fair slope. Even where the country has started as a plain, in a few million years the rivers bring down enough mud or gravel from the hills to produce a sloping countryside. When the hills are worn down we may get flat plains again, but there are quite enough hills left in England to prevent this happening for a very long time.

In the last million years this process was interrupted by the Ice Age, or more accurately a succession of four Ice Ages. Ice does not necessarily move downward, like water. On the contrary, it can be pushed some way up a hill by pressure from behind. It can also carry much more solids than water, including even large rocks. And it has done so over most of England north of the Thames. There should really be three geological maps of England. The existing one shows the underlying rocks, sands and clays. Another one, which is incomplete, would show the drift, or clays and sands deposited by ice. Finally a third, which has only been made in a few counties, would show the differences in soil, which are largely due to agriculture, and of course may vary greatly from one field to another. The glaciers which carried the drift mostly came from the north, both from the northern highlands of England, and from Scandinavia over the bed of the North Sea. In eastern England the most important drift is called the Great Chalky Boulder Clay, which contains fragments of chalk scraped off the Yorkshire and Lincolnshire Wolds, which were once real hills like the Chilterns and the North and South Downs.

Great glaciers interfere with drainage in four ways. They plough up the soil below them. They lay down beds of drift. And near their ends they make what are called moraines, that is to say banks of boulders, gravel and clay dropped as the ice melts. Finally they dam up valleys, forming lakes from which the water must find another way out. The vale of Pickering, in Yorkshire, used to drain into the sea. The ice from the North Sea dammed it up, and the water in Lake Pickering tore a gorge into the Ouse valley. So to-day the Derwent rises within less than a mile from the coast, and flows inland to swell the Humber. The fen district was a lake for quite a long time. The water could not get out through the ice-dammed Wash, and cut a valley through the East Anglian chalk, which is now dry, but once emptied into the Waveney. Meanwhile mud was laid down in the lake bed, which became extremely flat.

The flooded areas in the fens and in Yorkshire have all been lake beds at one time or another during the Ice Ages. When the ice melted they became marshes, and at times the fens have been covered by sea, as is shown by beds of sea shells. The level of the fens has risen partly by the laying down of mud, but partly through the formation of peat in swamps. This peat is not now much used as fuel, but it is of great interest because a microscopical examination of the pollen in it leads to a history of the English climate in the last ten thousand years. It was, for example, a good deal warmer than now six thousand years ago, as shown by the prevalence of pollen from lime trees.

During the middle ages monasteries were built on islands in the fens, such as Ely and Crowland, and the monks did a bit of drainage. The biggest drainage schemes were carried out in the seventeenth century when the bourgeoisie was gaining power. The great planner was a Dutch engineer called Vermuyden who was however somewhat grossly swindled by the Russell family, the ancestors of the Dukes of Bedford, the Earls of Ampthill, and the Earls Russell, who appropriated much of the drained land. But even if Vermuyden had got an earldom and a hundred square miles, I don't suppose the men who dug the dikes would have got much of the land that they retrieved. On the contrary, the inhabitants of the fens put up a fierce resistance when their small holdings were seized without compensation.

The fens were first used mainly for pasture, and in Cobbett's time, a little more than a century ago, they were full of cattle. Now they are some of our finest arable land. But the drainage of peat bogs is a very tricky matter, for this reason. When sand or clay dries, there is no great shrinkage. But when peat dries it contracts greatly, and moreover is gradually oxidized. So the land of the reclaimed fens falls slowly. Probably it was only the invention of the steam engine and its use for pumping which prevented large parts of them from going back to swamp. The present floods have doubtless laid down a little mud, but many centuries of flooding would be needed to raise the level so as to make pumping unnecessary.

What ought to be done about the fens? I don't know. But if I had the responsibility, the first thing I should do would be to consult the Dutch experts.* The Dutch have similar problems over half their country, and their lives depend on tackling them. They have tackled them on the very vastest scale. I recently went along the great dam which has been built across the mouth of the Zuyder Zee in a van belonging to *De Waarheid*, the Dutch equivalent of the *Daily Worker*. This dam is the most impressive of the works of man that I have seen, including the pyramids and the New York skyscrapers. In the middle of it we were out of sight of land. On one side was the salt ocean, on the other an artificial fresh lake most of which will gradually be drained as salt soaks out of the soil, and mud from the Rhine is laid down on the sand.

I expect that the Dutch would recommend more channels and straighter ones to carry off the water, as well as stouter banks, which are more easily made in England than Holland, as we have more stone.

When wars are a thing of the past, and men can plan constructively a generation or so ahead, I expect that a dam will be

---

* This has actually been done. I mention it because suggestions made in these articles are very rarely accepted.

built round the higher parts of the Dogger Bank in the North Sea, where there is an area as large as several English counties less than sixty feet below mean sea level, which could be drained and made fit for agriculture. It was land not so very long ago, as is shown by the peat dredged up from it, and it will be land again, when we are more interested in conquest from nature than from other people. But meanwhile we have the more urgent job of protecting the Fens and Yorkshire from floods. I hope we shall get on with it.

9

Pollen and History

A CONGRESS was held at Moscow University on the 25th anniversary of the first Russian study on pollen analysis as an aid to geology. It has been used for a variety of purposes, including the construction of an underground railway at Leningrad. Readers may well ask what on earth pollen has to do with underground railways, except that a tube may be a good refuge for a sufferer from hay fever during a month when there is a lot of pollen in the air.

The answer is curious. Though a pollen grain can only be seen with a microscope, its wall is not only remarkably tough, but resists decay much better than wood or leaves. So pollen grains are found in mud or peat thousands of years old where most other vegetable remains are unrecognisable. What is more, pollen grains have very characteristic shapes. That of the pine is quite unlike those of the broad-leaved trees. The oak pollen grain has three nicks which remind one of the nicks in its leaves, and so on. There is no difficulty in counting a thousand pollen grains in a small sample of mud, and assigning them to their correct trees, or, in many cases, herbs or grasses.

The pioneer in this work was a Swedish scientist called von Post, and Godwin has been his main follower in Britain. All over north-western Europe the first trees to grow after the last ice age were Scottish pines and birches. They were succeeded by beeches, oaks, and other trees which demand a warmer climate. And there was a period, probably about six thousand years ago, when the climate was much warmer than to-day. Pines grew 1,200 feet higher upon the Alps, hazels grew in northern Sweden two hundred miles nearer the North Pole than at present, and so on. Then the weather got a lot colder and wetter during the iron age and the Roman empire. It improved about A.D. 500 and then relapsed in the middle ages.

The whole story is extremely complicated, and naturally differs from place to place. For example in the British Fens there are layers of mud laid down continuously since the end of the Ice Age in some places. But at several levels, this mud contains no pollen, because the country was under salt water. This was so, for example, during the Roman occupation.

Pollen analysis may even help us to distinguish the truth or falsehood of various legends about Kings Arthur and Alfred. Godwin has recently been investigating turf which is being cut in the drained bogs round Glastonbury in Somersetshire. This is the region where, according to one tradition, King Arthur retreated to the isle of Avalon when wounded, and according to a much better one, King Alfred took refuge from the Danes at Athelney.

Godwin has already traced a number of changes of water level in this region, with their rough dates, and shown that one island in the marsh was connected with the rest of England by a causeway of brushwood bundles for a time, and later abandoned. But in the time of Kings Arthur and Alfred England was a good deal less rainy than now, whereas when most of the legends about them were written down, it was somewhat wetter. I do not know if it is certain that there was a marsh at Athelney during the Anglo-Saxon and Danish invasions.

Now most of Russia is pretty flat, and there are still bogs over a wide area in the north, and peat from former bogs over a much larger area. In particular Leningrad is built on marshy soil. So the opportunities for work are better than in England. Moreover, throughout the Union outdoor sciences such as geology and the study of soil have been developed very highly. In consequence more than a thousand papers on the subject have been written. The importance of pollen analysis for excavating tube railways is this. One can sink a borehole into the soil and get up a core with samples from different depths. A few hours' work with the microscope gives the date at which the mud was laid down, once the sequence of types of vegetation is known. Thus one can predict the type of soil to be met with. In London one usually soon gets down to the London clay, which was formed in salt water, and contains no pollen. But at

Leningrad pollen analysis was probably used to enable the tubes to be driven through those layers which contain least water.

According to the accounts of the pollen congress, Soviet workers are using the same method on coal seams. Some coal seams, for example the Better Bed in Yorkshire, are largely made up of spores of plants resembling the modern club-mosses and ferns. These plants did not bear seeds, but produced vast amounts of spores no larger than pollen grains. It seems that the Soviet workers have been able to classify them, so as to identify the same coal seam in different areas. In Britain this is more usually done by fossil molluscs in the harder rock between the coal seams. But if it can be done directly from the coal this is clearly an advantage.

One of the great developments in microscopic geology has been the use of the remains of single-celled animals which made hard shells to date limestone to assist in oil prospecting. Unfortunately many of the results are kept secret by the oil companies which use the information. Still another development is the discovery of the remains of simple plants like the threads of green and blue algae which grow in stagnant water in very old pre-Cambrian rocks formed before animals had evolved so far as to have recognizable shells or skeletons. In fact geologists are beginning to find the microscope as necessary a tool as biologists have done in the past. And its use may lead to as great advances in geology as it did in biology.

I O

Science and Prehistory

SEVERAL readers have asked me to write on the dawn of human history. I am not going to. I do not know enough about it. What I can write about is the impact of scientific method on archaeology. It is a curious fact that prehistory is in some ways more scientific than history. This is partly because so little was known about it before historians were affected by scientific method.

If you are trying to write the history of mediæval England you are up against the fact that all contemporary documents were written by priests or monks. So a king who gave large gifts to the Church was a good king, and one who took back his predecessors' gifts was a bad one, though if he got a lot of Englishmen killed in France or Palestine he might be glorious. Similarly a Roman emperor who killed a lot of senators was a bloodthirsty tyrant, while if he enslaved a thousand times as many Britons or Dacians he was a hero who extended the limits of civilization. One has to see through this before one can decide whether after all Nero and Macbeth were bad men. In fact Nero almost certainly was bad after his first five years. However, Macbeth had a good claim to the Scottish throne, and reigned peacefully for seventeen years, apart from one English invasion.

The student of prehistory has to judge men not by what other people said about them, but by the things they made. In fact every prehistorian must be something of a Marxist, because we only know of prehistoric men by their productions. So the first thing to do in establishing what happened is to be able to date an implement or a vessel, or at least to assign it to a period. Only occasionally can we establish a succession with certainty, as when a cave has been used by men over thousands of years, and the earlier deposits have not been disturbed. One of the very real difficulties is that many of the objects, such as weapons and ornaments, were buried with the dead.   This means that if a grave is deep, we may find weapons of a later period buried below those of an older one.   On the other hand a tomb, such as a long barrow, might be used for burials hundreds of years after its construction, as distinguished men are still buried in Westminster Abbey.

Only very careful work overcame these and similar difficulties. My cousin Lord Abercrombie devoted much of his life to determining the succession of pottery in Britain so that it was possible to date a site from a few fragments.   Many people found his work very boring, but such fragments play the same part in prehistory as do fossils in geology.   Of course the dates are not exactly known when the order of succession is known.   One can certainly describe a harpoon as Maglemosian or a vessel as Iron Age B, but one may be centuries out if one tries to give it an actual date in years, though one knows that the Iron Age was later than the Bronze Age, and so on.

A wholly new technique was introduced, mainly by O.G.S. Crawford, about thirty years ago.   He discovered that photography from aeroplanes would show up human work which was quite invisible from the ground.   For example an old road or wall on ploughed land may be visible as a strip where the ground is a little drier than elsewhere, and the wheat ripens a few days earlier.   A photograph just before harvest shows it up, while at other times of the year it is invisible.   Or a gentle rise in the ground marking an ancient mound may be clearly visible when there are long shadows at sunrise or sunset, and not at other times of day.

Among the discoveries made in this way is "Woodhenge" near Stonehenge, the site of an ancient wooden building nearly as large as Stonehenge, and of much the same design, probably built about 1800–1400 B.C.   Aeroplane photographs showed up the holes in which the posts had been fixed, and digging revealed traces of the timber.   Another similar temple was found at Arminghall, near Norwich.   Equally striking was the discovery of the road leading from Stonehenge to the river Avon.   Meanwhile geologists had found that while the larger stones of that great temple came from Salisbury Plain, some of the smaller ones came from the Prescelly Mountains in Pembrokeshire. As they weighed several tons, they must almost certainly have been brought by sea round Land's End, and up the Avon to the end of the road, thus proving that southern Britain was fairly peaceful over three thousand years ago.

Technology is an important part of the archaeologist's trade. Only after archaeologists had themselves learned how to make flint implements was it possible to classify the different types scientifically, according to the methods used in making them. To take another example, primitive pottery was shaped by hand, long before the potter's wheel was invented. The potters left their finger prints. And Soviet archaeologists believe that all the earliest pottery in their country was made by women, the first certain example of the division of labour.

The actual human remains do not generally tell us a great deal. The men of the later periods of the Old Stone Age, and of the Mesolithic, Neolithic, Bronze and Iron Ages which succeeded it, were all of modern type. It is an impressive fact that men have not changed much physically in the last twenty thousand years, though civilization has arisen. When we go back for hundreds of thousands of years we certainly find beings quite unlike any existing races, but who must be regarded as human, since they used tools, and often fire, though so far as we know they had no art, and their tools were very crude.

The evolution of domestic plants and animals has been enormously quicker than that of man, partly because men selected them deliberately, partly because the great change in their environment when they were domesticated subjected them to novel selective forces. In particular they are much more crowded than in nature, and have to acquire immunity against diseases due to overcrowding. So one can follow the evolution of the domestic dog, cow, or pig much more easily than that of man.

Perhaps the most remarkable of all scientific methods is the dating of ancient events by eclipses. The earliest date known is 2283 B.C. In this year the sun was totally eclipsed at Ur, in Iraq, and next year it was captured by the Elamites. The eclipse was supposed to have foretold the defeat. Homer in the

E

Odyssey records an eclipse of the sun when the hero Odysseus, or Ulysses, returned to his home in Ithaca after ten years campaigning and nearly as long staying with Circe and Calypso. The last eclipse there visible was in 1190 B.C., so Troy must have been taken about 1200 B.C.   These are some of the ways in which science is influencing the study of human prehistory.   I only wish it were being applied as directly to politics in our own country.

Was England a Yugoslav Colony?

SOME day we shall have a readable history of Britain which will tell us how the ordinary man and woman lived at different times. It is unlikely that there will be many names in it before the time of Caesar's invasion, which occurred just 2,000 years ago, though we do know the names of a few British "kings" before his time. Clearly there must have been a considerable political organization, and widespread peace, to allow the transport of large blocks of stone from South Wales to Stonehenge on Salisbury Plain. But all we know about the people of those days is the things they made, and their bones. In fact prehistory is the history of human production. Fortunately it turns out that if we know enough about production we know the most important things about a people. Their social organization depends in the long run on their daily work, and not the other way round.

One of the landmarks in English prehistory was the introduction of metals. The first metal used in Britain was bronze, and the first bronze objects which can be dated with any certainty are found in the graves of a people whom archaeologists have called the Beaker Folk, after the earthenware vessels which they made. The Beaker Folk, who had rounder heads than the earlier inhabitants, came to eastern England from Europe, especially from what is now Belgium. Another group of colonists, who introduced the custom of building with large stones, came northward along the Atlantic coast, but the Beaker Folk came from central Europe.

The route by which their culture spread has been traced largely by Yugoslav and Czech archaeologists. It came up the valley of the Danube. One of the main sites whose excavation has cleared up the story is at Vinca, a little below Belgrade, which was excavated by workers organized by Professor Vassits of Belgrade. Vinca was occupied long enough to leave about twenty-five feet of rubbish. The first inhabitants lived in holes in the ground walled with wattle, and probably roofed with hides. They used stone-bladed hoes for gardening and horn harpoons for fishing. But they made pottery which shows that their culture came from the south-east. For example the lids of their pots often had rough human faces. This type of lid is quite common in what is called Troy II, a city which was destroyed about 1800 B.C., and on whose ruins was built the city that Agamemnon and Achilles besieged about 1200 B.C.

As time went on Vinca continued to reflect the civilization of Troy. Later on its people had very beautiful images of a mother with a child, probably a goddess, and metal objects like those found in various parts of Greece. The main reason for this northward expansion was the existence of gold in Transylvania, and later the discovery of tin in Czechoslovakia, which was used to harden copper into bronze round the eastern Mediterranean. The Danube was gradually opened up for navigation, and a fair number of objects which are commonplace in the ruins of Troy II have been found in Czechoslovakia. They include bronze pins, earthenware mugs, earrings of bronze wire, and so on. Perhaps the most remarkable evidence of prehistoric trade in this region is the discovery in graves at Lengyel in Hungary, of ornaments made from the shell of a clam which lives in the Indian Ocean.

From Czechoslovakia these first users of bronze spread to the lower Rhine valley and into Belgium. And from Belgium some of them sailed or rowed to Britain. On their way through the forests of Germany they had lost a good deal of their culture. But they were enormously ahead of the original inhabitants. The Belgians whom they replaced had no domestic animals or plants, and were still chipping and not even polishing stone implements. This migration of culture was certainly not organized conquest, though no doubt there was fighting, as there always is when agriculturalists replace hunters. Britain two thousand years ago was not in any sense a Yugoslav colony. But we did get a very important element in our culture from Yugoslavia.

One reason why civilization spread from the Mediterranean to central Europe up the Danube rather than across the Alps was that four or five thousand years ago the Alpine glaciers and snowfields came down a good deal lower than they do now, so that it was almost impossible to get from Italy to Austria. Another reason was that along the Danube valley there are many patches of the fine soil called loess, which does not readily grow forests, but is easily cultivated, and forms for example the wheat-fields of Ukraine and Hungary. So men could row or sail up the river and start a primitive kind of wheat growing without having to clear the forest.

Finally there were the gold of Transylvania, the tin of Bohemia, and the copper of Slovakia. There was similar mineral wealth in Spain, which was early opened up for trade with the eastern Mediterranean, but not in France. Later on, when iron began to be smelted, iron ore was found in many places, and the Danube became much less important. Moreover, tin was discovered in Cornwall, and a trade route from Britain to the Mediterranean by way of western France was established.

Most of the facts of which I have written were discovered in the last thirty years. Doubtless many details will be filled in when archaeology starts up again in Yugoslavia. It is perhaps timely to mention these facts because we are all taught in school about the influence on British civilization of Greece, Rome, Palestine, and so on. Probably not one person in a thousand realises that we owe a very considerable debt to the countries of central Europe. It is likely that we are going to learn a good deal from the Czechs and the Yugoslavs in the next generation. The Czechs in particular are building socialism in a way which we may find easier to follow in some of its details than the Russian way.

Our first metallurgy came from Bohemia, though not of course directly. Our great English mediaeval reformer Wycliffe inspired the Czech Hus who in his turn influenced the British reformers of the sixteenth century, and perhaps once more we shall learn a lesson in civilization from the peoples of the Danube basin.

Cave Diving

THERE must be millions of young men and women in the Soviet Union who did not take part in the war either as soldiers or partisans, and feel the need to show, not only to others, but to themselves, that they are as brave as the heroes of the Revolution and the Patriotic war. They can do so by pioneering in the Arctic and other undeveloped regions of your great country. In a small and long settled land like England such opportunities are rare, but I will write about a small group which has found them, the Cave Diving Group, composed mainly of young engineers.

We have several regions in England where caves have been formed by the action of streams in limestone. One is in Yorkshire and Derbyshire, another in the Mendip Hills in Somersetshire. Most of the caves which are so far known are those which are no longer filled with water because the stream which formed them has dried up, is no longer in existence, or has found a channel at a lower level. From these one can often descend to the level of the present stream which appears out of one hole and descends into another. The submerged caves are being explored to-day. It is impossible to go far with a diving dress to which air is pumped through long tubes. So the divers use dresses of a type developed during the war for offensive operations. The divers carry enough oxygen compressed in small cylinders to last them for about two hours, or with some apparatus, five hours, during hard work, and much longer if they are at rest. I have been privileged to accompany them on visits to some areas already explored, though naturally a man fifty-five years old like myself is unsuited for the very arduous work of discovering new routes.

It is very impressive, after descending to the bottom of a pool about two metres deep, to crawl head-first down a narrow hole to a depth of four or five metres, and then to drag oneself with the hands through a narrow and low-roofed tunnel until one emerges to stand up in a great underwater chamber, stretching far beyond the range of our powerful electric lamps. The greatest difficulty is from mud. One cannot work when the stream is flowing very quick, and when it is flowing slowly the clouds of mud obscure the view down stream, though one can see well enough ahead.

This group has already obtained results of real value, including the discovery of the skulls of former cave dwellers, and a very remarkable vessel composed mainly of lead, probably used by them, but possibly dropped by a visitor at a later date. But their main task is surveying; and although they sometimes come up into hitherto unknown air-filled chambers, most of their surveying must be done under water, a difficult task. These surveys are of great scientific value. For there is little doubt that caves have been formed by the action of underground streams. These streams very possibly acted at a much greater depth than at present, and during an earlier cycle of erosion. Nevertheless, the air-filled parts of caves do not grow, and on the contrary are gradually filled up by stalactites. Whereas the submerged portions may at least be growing at present by erosion. By surveys made at intervals of some years it should be possible to find out whether streams are still at work hollowing out caves, and if so how fast. This may make it possible not only to decide between different theories of their origin, but to estimate how long their formation took, and thus to furnish a new method for dating the past.

I hope that, if they have not already done so, similar groups in your country will be able to acquire diving equipment no longer needed by your navy, and to explore the still growing parts of your country's caverns, thus helping to found a dynamical speleology. If such groups exist already, some of the members of the Cave Diving Group send their fraternal greetings, and hope to exchange information with them.

III
THE WEATHER

I

Frost

WHY are frosts so important? Why does it matter more when the temperature falls below 0° Centigrade or 32° Fahrenheit, than when it falls below 20° or −20°? The answer is because water is the commonest substance which undergoes a change of state within the temperature range where human life is possible. If we could live at such temperatures, the boiling point of water, or the temperatures at which the commoner kinds of stone melt, would be equally important.

By a change of state is meant a change between solid and liquid, or liquid and gas, more rarely a change between solid and gas, skipping the liquid state. Most pure substances change very abruptly from one state to another. The freezing point however does not depend only on temperature, but on pressure. Most liquids get denser when they freeze. For example when one melts lead or solder the solid unmelted part remains at the bottom of the pot, though there is a thin skin at the top where the surface is cooled by the air. In consequence if you squeeze molten lead you tend to force it back into the solid form. In other words the melting point of lead rises with pressure.

But water behaves differently. Ice is lighter than water. That is to say water expands when it freezes. It even starts to expand a little before it freezes. This has a number of consequences. If ice were heavier than water it would fall to the bottom of ponds and of the sea. Shallow ponds would freeze solid. So would the Arctic Ocean. And very likely there would be ice at the bottom of even the tropical oceans. In fact the whole earth would probably be a lot colder; and perhaps Venus, which is a good deal warmer than the earth, would be a more suitable place for living beings.

On the other hand frosts would be less dangerous to traffic. The main reason why we slide on ice is apparently that local pressure causes it to melt, so it does not give us as solid a footing as an equally smooth surface of wood or metal.

It is not so sure whether the other menace of frost to town-dwellers, namely burst pipes, would be any less serious. When the water freezes in a pipe, it bursts it by expanding, particularly if some water is trapped between two blocks of ice. We do not notice the burst until the thaw occurs. But we may be able to prevent it, not only by leaving taps dripping, but by heating up a frozen pipe the moment the water stops flowing, if we can find out where the ice has formed. If water contracted on freezing, the pipes would not burst till it thawed again, but they might do so then. I leave that question to better physicists than I.

It is not only water pipes that are smashed by frost. Wet ground, and even wet rocks, are split. The sharp-edged pebbles so produced in mountains and the Arctic are quite characteristic. So is a peculiar type of soil produced when the stones and earth on a hillside have been loosened by frost, and the soil frozen so deep that when spring comes and the surface melts the water cannot sink into it and the surface layer slides down. Geologists can recognize this type of deposit, which is called Coombe rock in southern England, where it was formed during the Ice Ages, and accumulated in valleys in the downs just beyond the range of the great ice sheets which covered much of the country.

When waterlogged soil, such as a Siberian tundra, freezes, it first expands, but with severe frost it contracts again, and cracks form, often with a loud bang. These fill with ice which does not melt next spring, and the cracks may grow till they are two yards across.

Frost also forms peculiar rings of stones on the surface. Apparently heat leaks out quicker through a solid stone than through soil, so ice forms under stones and lifts them to the surface. In the Arctic there are no earthworms to counteract this process. So stones are heaved up above the soil, and a further action of the frost, about which there is a good deal of dispute, arranges them in rings. These, and other peculiar types of soil due to frost, have been found buried below the ordinary soil in many European countries. During the ice age it seems that England and France had plenty of snow in winter, but few frosts hard enough to make deep cracks. While in central and eastern Europe conditions were more like those in Alaska to-day.

The most serious effect of frost is of course on living beings. They all consist mostly of water, and when this freezes they are at best in great danger. Nevertheless, some can stand freezing. Of course a fish in a block of ice is not necessarily frozen. For living stuff contains enough salts and other things dissolved in it to lower its freezing point well below that of fresh water.

Curiously enough those simple organisms, such as some plants and bacteria, which will survive freezing, are safest if they are made extremely cold, in fact brought down to liquid air temperatures. At temperatures a little below freezing point the ice molecules can still move a little, and arrange themselves in large crystals which destroy the organic structure. At very low temperatures crystals can no more grow in ice than they can in glass at ordinary temperatures.

Most of our plants have by now shed their leaves, and those which have not have special chemicals to protect them. In fact most leaves of winter evergreens such as holly and spruce are full of oil and resin which will not freeze, and contain very little water. But unfortunately these frost-resistant liquids will burn. That is why Christmas trees and holly branches can rather easily catch alight, and it is safer to decorate them with electric bulbs than with candles. I suppose that every winter these anti-freeze mixtures in leaves are responsible for a few human deaths, mostly of children, from burning. The anti-freeze compounds used in car radiators also kill a few people who drink them, particularly when they are used for adulterating cheap spirits.

In England frost is not much more than a nuisance. In the northern parts of the Soviet Union it is one of the main natural conditions to be fought. Even the principles of architecture are quite different where in a few winters' frost can heave the foundations of a house out of the ground. The year is sharply separated into two parts, during which people have very different kinds of life. A Soviet town planner must think of both, and the conquest of the Arctic so that ordinary people can be as useful and happy in winter as in summer is one of the many problems which the Soviet Union is solving.

2

Can we prevent Cold Spells?

I HOPE that by the time this article is printed*, the cold spell will be over. But I am writing on the assumption that it is still going on. The business of a scientist is first to describe what is happening, and then to suggest what should be done about it. I shall try to do this.

Weather is a matter of eddies. The air of our planet forms a relatively thin film, and air can only move into one part of it if most of the air already there moves away. Some of the eddies are very large, for example those of which the trade winds form a part, others are only a few hundred miles across, and of course still others very much smaller. When cold air descends from above to the ground, it must spread out. If it goes north in the northern hemisphere it keeps some of the eastward motion appropriate to the latitude from which it started, and therefore swings eastwards. So a patch of heavy cold air descending and spreading outwards generates a clockwise eddy or anti-cyclone. A patch of light air tends to rise, and generate a counter-clockwise eddy or cyclone, but in rising it cools down, loses its water vapour as rain or snow, and thus becomes relatively denser. That is one reason why cyclones are much less persistent than anti-cyclones.

Unfortunately for us an anti-cyclone, or "high", that is to say an area of high pressure, or overcrowding, in the air, has developed over southern Scandinavia. The air is streaming out from it in clockwise spirals, and in consequence we are getting east winds from frozen Europe. There is plenty of warm air out in the Atlantic, but it cannot penetrate this persistent eddy. I do not know how the cold spell will come to an end. A very likely way is that warm wet air from the Atlantic will force its way over the cold eddy. If so there will be terrific snowstorms

* February, 1947.

along the front where the warm and cold masses of air meet. So the frost may end in a violent snowstorm followed by a thaw.

If the cold spell could have been predicted even a month ahead, the government could have taken measures in advance to build up coal reserves. In some countries, but not yet in Britain, such prediction is possible. For example predictions can be made six months or more ahead about the Indian monsoon. They are not always right, but they are correct enough to be well worth making. However the monsoon is part of an eddy which lasts for some months, whereas our cyclones last a few days. Meteorology has got to the stage of predicting about one eddy ahead, but not much further.

No doubt weather prediction will be improved, but this will be a slow process. Until recently there was no possible way of influencing the weather, simply because it would take many million horse-power to reverse the flow of wind over any large area. Nevertheless I think it just possible that we could have alleviated this cold spell considerably if we had tried. G. N. Lewis showed that a few tiny crystals of solid carbon dioxide, or "dry ice", dropped into a vessel containing air saturated with water vapour, will cause it to condense; and if it is cold enough, will cause ice crystals to form.

This is an impressive experiment in the laboratory, but it also works out of doors. By spraying dry ice from an aeroplane into a cloud, a local snowstorm has been caused. The amount of dry ice needed is a very tiny fraction of the weight of snow produced. A snow crystal has to be of a certain minimum size before it shows any strong tendency to grow. And the dry ice crystals are enough to start off this process in clouds which are in an unstable state and ready to condense into snow.

If large numbers of aeroplanes flew over the North Sea, spraying the clouds with powdered dry ice, it would be possible to provoke snowstorms in the clouds which are blowing towards us from the east. This would have two effects. The condensation of water to ice would warm the air, and if the clouds were cleared away we should have sunshine in England. On the other hand when the operation ceased at night we should have clouds again, which would blanket the country, as they do at present, and prevent severe night frosts.

Such an operation would demand the use of every available aeroplane. It would require the production of many thousands tons of dry ice. And it would need special apparatus for dispersing it from the aeroplanes in the form of fine powder. These things could not be improvised in a week, perhaps not even in a month. Everything needed could easily be made in a year, and the whole thing would cost less money, let alone lives, than a single night's raids on Germany. It is probable that the Air Force at present stationed in Britain is not enough to cut a swathe in the clouds big enough to protect the whole country. It could almost certainly protect Greater London, or the main urban areas of Yorkshire and Lancashire. The mere gain in daylight and warmth would mean a great saving of fuel.

There are probably a few other meteorological situations where it would be possible to provoke rainfall or snowfall in one area rather than another, thus bringing needed rain or averting floods or blizzards. The conditions are that there should be clouds just ready to condense into rain or snow, and that the wind speed should be slow, so that a few hundred planes could deal with the clouds coming in over a fair stretch of coast.

I believe that the R.A.F. should be encouraged to attempt operations of this kind for two reasons. The first is to help them to realise that they are public servants like the postmen or teachers, and to help the public to feel towards them in peace as we felt about the fighter crews in war. The second is perhaps equally important. In peacetime the R.A.F. carries out various sham operations. The crews concerned are told that a particular operation would or would not have been successful. But the whole thing tends to be rather artificial, like army manoeuvres. It is a great thing for a fighting service to carry out an operation whose success or failure they can see, even if no enemies are killed. And it is a better test of efficiency than most practice operations.

During the Russian famine of 1921 Lenin set up research to prevent the next famine. The main thing to do to avert another situation like the present is to accumulate coal reserves, which means more miners, and better equipped mines. But we should also begin serious work on weather control. And such work, if only by two or three aeroplanes, could begin next week.*

* In fact, the first experiment of this kind over England, by a single aeroplane, was made in 1949. At least one other substance has since been found to be more efficacious than dry ice.

F

3

Weather Forecasts

THE Air Ministry predicted a thaw in southern England beginning on March 6th, 1947. Instead we got a heavy snowstorm, and had to wait for the thaw in the London area. I think the Press was quite unduly polite about this episode. The general public was not. It blames the scientists concerned. It may be right to do so. Certainly someone is to blame. It is not my business to find out who. That is the business of the House of Commons.

Reports of air pressure, temperature, wind speed and direction, and cloud, rain, and so on, are constantly being received from a number of stations at ground level, and from ships at sea, by an official attached to the Air Ministry. In addition they probably get some reports from aeroplane crews as to the state of the upper air, and also reports of the movement of small pilot balloons sent up. They may use radar, which, among other things, can detect clouds at a great distance.

On this basis they make up the maps of pressure distribution and winds which were familiar enough before the war. These maps also include the position of "fronts" where cold and warm bodies of air meet at ground level. Rain and snow are formed along such fronts when the warm air is wet, and though this happens above ground level, and the boundaries between the air masses are far from vertical, but slope a lot, the position of the front at ground level helps prediction greatly.

One or more meteorologists then make a forecast on the basis of this map, and of their knowledge of how different configurations of air have changed in the past. Most meteorologists find prediction a hateful job. It must be done at top speed, whereas the best scientific work is rather slow. I always like to leave a scientific paper about for some months, and then to go through it, looking for mistakes. Above all, forecasting is always uncertain, and sometimes very uncertain. No scientific prediction is one hundred per cent. certain. The most certain predictions are those of the directions of the sun, moon and stars published in the nautical almanac. But the moon may be several seconds early or late at an eclipse, and it is probable that some time in the next few million years a body from outside will come near enough to the solar system to throw the planets appreciably off their predicted paths.

A meteorologist can make certain predictions with over ninety per cent. certainty. That is to say he would make money if he consistently betted nine to one on their being right. The predictions of rain a day ahead when a cyclone is approaching from the Atlantic in normal weather are of this kind. Something very like it has happened so often before that there is little doubt. On other occasions the probability of success is only about two to one, or even less. This may happen for two different reasons. There may be a fairly familiar situation, but it may be doubtful whether a front with snowstorms will or will not pass over London. Or the situation may be almost without precedent.

This is the case at present, in 1947. There has certainly been no spring like the present one during this century, and though there have been a number in the last five hundred years, none of them were properly reported from the meteorological point of view. What actually happened last week was this. A cyclone, or counter-clockwise eddy of warm, moist air was approaching England from the Atlantic. In normal conditions it would have pushed the cold air covering southern England out of its way. Actually it failed to do so.

The scientific way of publishing forecasts would be something like this. "There will be snowstorms in south-eastern England. (P > ·9). They will be followed by a thaw (P= ·5 — ·7)." Here P is an estimate of the probability of the prediction, and > means "greater than". This kind of prediction is often all that we can make in genetics. If two normal parents have had a child with a combination of blindness and mental defect of the type called amaurotic idiocy, we can't say what the next child will be. We can say it will be an idiot with the probability of ¼. In twenty such families we can say that there is only one chance in 317 that all the twenty next children will be normal.

In fact one can predict with high probability that there will be at least one idiot of this kind.

Now unfortunately neither the general public nor the civil service is accustomed to this kind of prediction, though by now the higher ranks of the R.A.F. ought to be so. In consequence some highly placed people demand impossibilities from their experts. Several things may have happened. A meteorologist may have made a genuine mistake, and predicted a thaw confidently. If so it might be advisable to replace him. He may have made a guess. Certain technical experts rose to high positions during the war by always giving confident answers to questions put to them, when better men hedged. They were generally right; and when they were wrong they managed to throw the blame on someone else. If there are such people employed on weather prediction it is time they were given less responsible posts.

It is equally likely that the unfortunate meteorologists have orders from a superior with no scientific qualifications to give as definite prediction as they can, and never mind if they are sometimes wrong. Such an attitude is partly due to contempt for the public, who are erroneously supposed to be so stupid that they cannot understand the notion of probability. Again the meteorologists' report may have been "simplified" for the press and radio, leaving out qualifications.

Finally there is a possibility of sabotage. People exist who are so profoundly convinced that socialism is wrong that they will do anything to make the present government unpopular. Such people are deliberately wasting current at the present time. It is unlikely, but not impossible, that sabotage has occurred here.

The only people who have a real opportunity of finding out what happened are members of parliament. They can and should ask questions, and may be able to find out the truth. The whole affair is but one more example of the extreme difficulties which arise when scientists have to work under superiors without scientific training. The way out is not, of course, to establish a dictatorship of scientists, who would be no better than any other dictators, but to see that the general public, including administrators, get some scientific training.

In the meanwhile, I hope that the Meteorological Department of the Air Ministry will be given a shaking up. In my experience such departments are apt to consist of a lot of hard-working subordinates and some rather complacent old gentlemen with a knack of getting on the right side of high officials. This kind of organization makes for smooth running of offices, but it does not make for scientific efficiency.

4

Heat and History

AT the time of writing (May, 1949), we are getting some slightly warmer weather, and may be going to have a bit of real summer. I should like to see some steady rain for a couple of days, followed by bright sunshine and hot weather for the Whitsun holidays. Unfortunately we cannot control weather as yet, and cannot predict it far enough ahead to arrange our holidays in a bright period. When I say we cannot control it, I simply refer to the possibilities under our present economic and political system. If the clouds are fairly thick, it is often possible to produce showers spreading powdered solid carbon dioxide or "dry ice" from aeroplanes. This was first done over the United States, and first applied on a practical scale over the Soviet Union. We could probably do something to help our harvest in the dry areas, and if we had enough rain, to get clouds to rain over the Irish Sea rather than on holidaymakers. It might cost as much as the Berlin air lift, but it could be done. Meanwhile we must trust to luck, and hope that on Whit Monday we shall be in an area of high pressure. This generally brings a fairly cloudless sky, with sharp frosts in winter and bright sun in summer.

Most men and women like warm weather, but it is the so-called cold-blooded animals, and especially the insects, which are most affected by it. They live quicker when it is warm. The easiest way to prove this is by timing ants with a stop-watch. If you find a track along which ants are walking fairly regularly, and time twenty or so of them over a yard so as to get an average you will find that the warmer the air near the soil, the quicker they walk. Roughly speaking, a rise of 13 degrees Fahrenheit will double their speed; and the American astronomer Shapley was able to measure the temperature correct to one degree by timing about 20 ants. If the ants have any consciousness they probably don't think they are going any quicker in the heat.

At any rate human beings don't.   If you warm up a man's brain
artificially or if it is warmed up by fever he thinks time is passing
more slowly.   That is to say if you ask him to tell you when a
minute is passed, he will tell you it has done so after only forty
seconds.   This is not because he is consciously or unconsciously
counting heart beats.   You can speed up his pulse rate without
altering his time sense.   It is simply because chemical processes
in his brain are going on quicker.

Thirty years ago the speed-up of insect life in London by hot
weather always brought on an outbreak of summer diarrhoea
which killed a number of babies.   This was due to infection of
their food by bacteria carried by flies.   To-day this is much less
serious, partly because mothers have learned to cover food up,
but mainly because, owing to motor traffic, there is less horse
dung in the streets in which fly maggots can grow.   Although
motor vehicles kill a number of children every week, they
probably save the lives of still more by keeping the streets
relatively clean.

Plants also grow quicker in hot countries, provided they
can get water.   Probably this is why most human migrations
have been towards warmer lands.   In Europe the human stream
has on the whole flowed westwards and southwards.   In consequence the Scandinavian countries are the only ones which
have never been conquered by foreigners, though they have
sometimes conquered one another.

Now unfortunately the warmer countries are more favourable
not only to plant life, but to insects and worms.   The insects
not only bite us, but infect us with diseases such as malaria,
infantile diarrhoea, and kala azar.   The worms include a number
which invade our bodies.   In consequence human life is generally
longer and healthier in cold than in hot climates.   It was a great
surprise to statisticians when, in the eighteenth century, they
discovered that Swedes lived about twice as long as Italians.

We now know that this need not be so.   The key to a healthy
life in a hot climate is insect control.   This can only be carried
out by public effort.   It is not much use keeping my water tank
free of mosquito larvae if they breed in my neighbour's.

One reason why the new state of Israel is to be welcomed

is that it will certainly develop public hygiene in such a way as to be an example to its neighbours. It includes men such as Adler, who has worked out how kala azar, a disease originally detected 'in India, is spread by insects in Israel and surrounding countries. His fellow-countrymen will certainly listen to him when he tells them how to avoid it. There are of course political reasons for the military success of Israel in the recent fighting. But a very important reason was certainly the higher hygienic standards of the people of Israel. When their neighbours realise this fact they will perhaps try to get the death-rates in Cairo and Damascus down to the level of Tel-Aviv, which will not be easy without a considerable redistribution of wealth and a considerable increase in education.

It is quite probable that in the next few centuries we shall see a reversal of historical trends. Four thousand years ago Europe was barbarous, but Egypt and Iraq were civilized. Pakistan was another great centre of civilization. If these very fertile but at present unhealthy countries can be made healthy again they may well pick up the lead which they lost. It is hard to see how they will do so without adopting socialism. Certainly they will need a social hygiene service on a bigger scale than our own, and I do not see how this could be organized under anything like their present economic systems.

One reason why the British officials in the Near East were mostly so hostile to Israel was of course that Labour was highly organized in Palestine and collective farms were proving successful, which put subversive ideas into the heads of a number of Arabs. Israel has still a long way to go on the economic path, but it is far ahead of its neighbours. First perhaps in the field of hygiene, but also in that of economic and social progress, it may once more be a guide to other peoples, as it has been in the past.

5

Autumn

THE leaves are beginning to fall, and soon most of our trees will be bare. Annual plants are dead or dying. and so are many animals which only live for a year or less. Others are preparing for winter in different ways. Many birds have already flown south. Life, in fact, is beginning to go on at a lower level of activity. We take this sort of thing for granted in Britain. But it is a shock to people from warmer countries. A cousin of mine from New Zealand, on landing in England in winter, thought we must be a decadent old country, as we didn't even trouble to cut down our dead trees.

And it is important to remember that an annual cycle is a special adaptation. Life is in some ways easier for plants and animals where there are no great changes of warmth and light. The different kinds of plants and animals which live in countries with a cold frosty winter have adapted themselves to it in different ways. The winter is difficult for two distinct reasons. Living substance is generally killed by freezing, largely because ice crystals grow in it and break up its structure. And owing to the shortage of light in winter even evergreen plants cannot make much new substance, so the supply of food for animals is cut down very greatly.

The simplest way of coping with winter is to die in autumn, after leaving a number of eggs or seeds which will not be damaged by frost, and will start the species up again next spring. This is the course taken by many of the smaller plants and animals. It involves special adaptations. In particular many seeds and eggs will not germinate or hatch till they have been exposed to cold, and then warmed up again. This acts as a safeguard against germination in autumn.

The deciduous trees shed their leaves, and in a great many plants all the parts above ground die off. Before the leaves are shed, most of the living substance is withdrawn from them into the branches and stem. In particular the green chlorophyll is broken down into constituents which can be used again next spring. If the leaves are killed by frost before this process is complete, the tree loses a lot of valuable materials. The signal which makes the trees drop their leaves is usually the shortening of the day. It used to be thought that walnut trees would not grow in Leningrad because the first frosts always came before they had shed their leaves. They can however be grown if a tarpaulin is pulled over them daily some hours before sunset for a fortnight or so before the first frosts are expected, so that the length of day is effectively shortened. If so they shed their leaves in time, and make other preparations for winter.

The evergreens protect their leaves from the action of frost in various ways. The most obvious is to fill them with drops of oil and resin which will not freeze. This is the method used by the holly and spruce, and is why holly leaves and spruce needles burn so nicely.

A lot of birds fly to warmer countries in winter. So do a very few butterflies. This means that they can exploit one area in summer and another in winter. No other animals undertake such great annual migrations. Man could certainly learn from the birds. In a properly organized society every factory and mine worker who wanted could get a month's work in the country in summer apart from an annual holiday. This would of course require far better housing in rural areas. One of the pleasant sides of an agricultural worker's life is that he doesn't do the same work all the year round. In a society planned by workers such variety would be normal.

Most animals eat less in the winter, and put on fat in the autumn, which disappears during the cold months. It is used up much less quickly in a cold-blooded animal such as a frog or a snake, than in a warm-blooded one such as a field-mouse or a weasel, which has to use food or fat to produce heat during the winter. However, the cold-blooded animals are liable to be frozen to death if they do not burrow deep into the ground or get into water too deep to freeze. Some warm-blooded animals, such as the hedgehog, compromise by going into a winter sleep in which their temperature falls, but not all the way to freezing point.

The animals which stay active through the winter mostly eat grass or other animals, and often grow thicker coats. Tree-leaves are mainly eaten by insects, which pass the winter in a dormant condition or as an egg or pupa. However a few animals make stores of food. Squirrels and some other rodents store nuts and seeds for the winter. Ants and bees also store food. Again a fair number of warm-blooded animals protect themselves from the cold by burrowing. They mostly use their burrows in summer too, but clearly a burrow is a very good protection from frost.

The yearly cycle is of course not the only one which plants and animals have to face. All except those which live underground or in the deep sea have to cope with the daily round of changes in light and warmth. Some sea animals have daily movements up and down which can be compared with birds' yearly migration.

Perhaps the most striking rhythmical changes in environment are those produced by the tides. The plants and animals which live in the tidal zone must be able to stand not only the violence of the waves, but drying up twice daily, not to mention frost and rain, unless they live in rockpools or burrow into sand or mud. The most severe conditions of all are those in the estuaries of rivers where the water changes from fresh to salt, and back again, in a few hours. Very few animals can stand up to this. Those few which are adapted to it are often found in immense numbers. Thus aquarists who go down to the Thames in London at low water can get very large numbers of the blood-worm *Tubifex* in a short time; in fact they spend longer in washing the mud off them before giving them to their fish than in collecting them.

Considered as an animal, man has coped with winter by storing food and burrowing from his earliest days. Later on he discovered fire and dressed in the skins of other animals. Still later he made clothes from fibres and built houses above ground. Storing and burrowing are not production, though they are steps in that direction. Social insects such as bees and wasps which do not merely store food, but make elaborate combs in

which to keep it, may be said to have started production.   A
hive of bees which can live through the winter because of their
productive activities, while most other insects die, is getting
rather near the human stage.   In fact man adapts himself to the
coming of winter in many more ways than any animal.   But it
is well worth watching how the animals do it, and even taking
hints from some of them.

6

Freezing and Thawing

DURING the last month freezing and thawing have become matters of immediate interest to all of us. They are examples of the abrupt changes of quality which are fairly common in physics.

There are gradual changes as well. As we cool down water from its boiling point it becomes gradually denser till it reaches a maximum density at 4 degrees Centigrade. Then its volume increases by about one part in 7,600 until it reaches freezing point, but as it freezes it expands by nine per cent. When further cooled ice contracts again, but never gets as dense as water. The expansion due to freezing bursts our pipes, though we do not notice this until the ice melts again, and water leaks out. Water is unusual in this respect. Most substances contract when they freeze, but the way in which water molecules are packed in ice crystals is rather uneconomical of space. Metallurgy would be a good deal easier if all metals behaved like water. But iron contracts considerably when it solidifies, so castings seldom fit their moulds quite exactly. On the other hand if ice were denser than water, it would sink to the bottom of rivers, lakes, and seas on forming, and would not easily melt in spring. So the world would be a good deal colder.

Physicists find it very hard to explain why most substances melt at a very exactly defined temperature, and do not behave like candle-grease, for example, which softens gradually. Melting is one example of a change from order to disorder. In an ice crystal each molecule has its place, in water it wanders about. Another similar change is demagnetization by heat. Each iron atom acts as a little magnet, and if they are arranged in an orderly way a whole block acts as a magnet. But this type of order disappears on heating before the iron melts.

Physicists have given pretty complete theories of matter in the gaseous and crystalline states. That is to say from a knowledge of the forces between atoms they can calculate fairly accurately how a gas or a crystal will behave. But they have not yet got a satisfactory theory of liquids. We know, as a matter of fact, how they do behave, and can make the necessary calculations about the flow of water, oil, or mercury, through pipes, or the motion of a ship in water. But we cannot yet show why a particular liquid has just the properties that it has. This is because a crystal is an almost perfectly orderly arrangement of atoms, and a gas an almost perfectly disorderly one, but a liquid is something in between. Similarly steel balls can be packed in an orderly way, or scattered at random, but when they are rolling over one another they are neither quite orderly or quite disorderly.

Economic theory is in a similar position. The "classical" economic theory describes what would happen if thousands of producers competed with one another in a disorderly way, and thus determined the price of commodities. This theory was nearly true a century ago. One can also describe the economics of a socialist community, in which production is orderly, being controlled so as to give the maximum use value from the labour expended, instead of the maximum profit. But our present economics are half-way between order and disorder. The state controls prices, and enforces the production of utility goods. But the producing firms try to make as large profits as they can. And economic theorists are hard put to it to give a satisfactory account of such a system. In fact it is a very unstable one. It will be up to us, at the next general election, to decide whether it is to develop in the direction of Socialism or Fascism.

Chemical disorder favours the liquid state at the expense of the solid and gaseous states. Mixtures are on the whole harder to freeze or boil than pure substances, that is to say they stay liquid over a greater range of temperature. For example the freezing point of water containing 3 per cent. of common salt is about $1\frac{3}{4}$ degrees centigrade below that of pure water, and the boiling point about half a degree higher. So if you throw salt onto snow which is just at freezing point it will melt it, and if you add salt to boiling water it will allow it to get hotter without boiling off, and thus cook potatoes more efficiently.

Another curious feature of the liquid state is that a liquid can be supercooled or superheated. One can sometimes cool down water several degrees below freezing point without its freezing. Such water may start freezing for no obvious reason. But it will certainly do so if the tiniest crystal of ice is dropped into it. The temperature at once rises to freezing point, as the formation of ice releases some energy. Similarly if water is boiled for a long time in a smooth vessel, so as to drive out all the air, and to provide no point where steam bubbles form easily, it is often possible to heat it above boiling point. When it boils there is something of an explosion. On the other hand it is quite impossible to heat ice above its melting point, except under very high pressure.

A supercooled or superheated liquid is in what chemists call a metastable state, analogous to a revolutionary situation in politics. A small amount of ice will organize the supercooled water round it into crystals, just as a few determined revolutionaries will organize a mass of discontented people to take definite action to end their wrongs.

A great many substances can exist either as solid, liquid, or vapour; as H_2O can exist as ice, water, or steam. Some cannot, for various reasons. Solid carbon dioxide is quite well known (in peacetime) as "dry ice". If you warm it, it goes off into gas without melting. It can exist as a liquid, but only under a pressure of five atmospheres or more. I am one of the very few people who have ever seen liquid carbon dioxide except through thick glass, and even held it in my hand, as I happen to have opened cylinders of the liquid while working in air at a pressure of ten atmospheres, and seen it flow out. Other substances, such as sugar, cannot be melted without changing their chemical composition. Others again explode before they melt. On the whole the harder a solid, the higher its melting point, because the same forces which hold the atoms together against ordinary strains prevent them from moving off into a liquid. So melting points vary from about minus 272 degrees centigrade for helium to three or four thousand degrees for carbon. Only in a few cases do the solid and liquid have different names, such as ice and water, fat and oil. But if we realize that all liquids will freeze, and most solids will melt, it helps us to think dialectically about nature.

IV

THE EVIDENCE FOR
ASTRONOMICAL THEORY

G

The Earth Spins

IN the series of articles of which this is the first, I am going to say something about the observed facts behind astronomical theories. We are taught in school that the earth goes round the sun, that the light from the nearest star takes several years to get here, and so on. We ought not to take such statements on authority. We ought to know something of the evidence on which they are based, and even to check it if we have the opportunity. Otherwise we shall get into the very dangerous habit of believing any story that we hear often enough.

I am not going to go into the evidence that the earth is round. The fact that maps are good guides to action, and are made on the basis of its roundness, should be sufficient. It would make all the difference to world politics if the earth were flat. It would mean that there were new regions to be found, and that ambitious statesmen could annex them instead of invading their neighbours. But it is not so obvious that the earth spins round once a day, or what difference it would make if it did not. Until recently people thought that the sun, moon, and stars spun round the earth once a day. The stars were supposed to be stuck to a solid object called a firmament. Children learn about the firmament in Bible classes, and unlearn it in Science classes. Centrifugal force would produce a fearful strain in the firmament, but perhaps if there were a firmament it could stand it. And when Copernicus argued that the earth went round and the stars did not, he could only urge that this view made astronomy easier to understand.

The first good bit of new evidence for the theory arose when it was shown that pendulum clocks went slow in the tropics, and that there were more yards in a degree of longitude near the equator than in France, more in France than in the Arctic. That is to say the smallest distance between two points on the earth's surface such that the maximum "height" of a particular star above the horizon is one degree higher at one than the other increases as you go to the equator. This is at once explained, and what is more explained quantitatively, if the earth's spin makes its equator bulge out. The combination of centrifugal force and greater distance from the centre lessens the force of gravity, and slows down a clock regulated by a pendulum, but not one regulated by a hairspring. A turning firmament might have some effects of this kind, but it would be very strange that it should have just the effects calculated if the earth turns once daily.

Another striking confirmation came from the study of winds. Air moving away from the North Pole has no motion due to the earth's spin. So as it goes southward the eastward-moving earth leaves it behind, and from being a north wind it becomes a north-east wind. In fact it turns right. The cold air descending in an anticyclone in the northern hemisphere moves right as it spreads out, so the anticyclone or "high" turns clockwise, and a cyclone with warm rising air turns anti-clockwise. The opposite is true in the southern hemisphere. Ocean currents and ice behave in the same way. A north wind in the Arctic Ocean drives the ice-floes south-west.

Probably the most dramatic of all the proofs of the earth's rotation was the experiment which Foucault made a hundred years ago with a very long pendulum suspended from the dome of the Pantheon in Paris. The pendulum, which was merely a weight on the end of a rope, was set swinging and left. Common sense suggests that so far as possible the pendulum will keep swinging in the same plane, or at least in a series of parallel planes, and physical theory supports common sense. The question is whether "the same plane" is fixed relative to the earth or relative to the fixed stars. When the experiment is done (and you can see it repeated any day at the Science Museum in London) the plane of swing moves relatively to the building, but keeps as steady as it can relative to the stars.

So it can be used as a clock like a sundial, though actually it is not so accurate. On the other hand the gyro compass, which is based on practically the same principle, namely that a spinning body given the necessary freedom will keep its axis pointing in the same direction, is very accurate, and of course very useful in aeroplanes. It is most remarkable that nobody did the pendulum experiment before 1851. It involves no mathematics until one tries to calculate the small allowance to be made for friction. All it needs is a large building with no serious air draughts, so that the pendulum can go on swinging for some hours.

A much more difficult experiment is this. A pair of metal weights on the end of hinged arms is suspended from a thread or wire. At first the arms are held out so that the weights are as far apart as possible. Then a thread which holds them up is cut or burned through. They fall until the arms are hanging vertically. In consequence the whole system begins to turn in the same direction as the earth, that is to say opposite to the sun's apparent motion. This is due to the conservation of spin, or angular momentum. The system is turning once a day with the earth. When the weights drop, the amount of spin remains the same, but it has a system with less moment of inertia to move, since the whole mass of the system is now concentrated near its axis. So it turns more quickly. The principle involved is the same as that of a flywheel. The mass of a flywheel is as far away from its axis as possible. So for a given number of rotations per minute it has a great deal more spin than if it were near the axis. It is harder to stop it or to set it moving.

Finally if Blackett is right in his recent guess that all spinning bodies are magnets, whose strength is given by a law which he states, the earth's magnetism is a further bit of evidence that it is spinning.

If we lived in caves, and had never seen the sun, moon or stars, but had made the other necessary observations, the hypothesis of the earth's rotation would probably be something which most scientists believed, but which one could doubt without raising any suggestion that one was mentally abnormal. If some of these caves communicated with the sea, so that the tides could be observed, a few daring thinkers would probably have deduced that there were one or two heavy bodies outside the earth. But they would not have calculated their sizes or distances. A sufficiently perverse and ingenious believer in a fixed earth and a moving firmament could perhaps have produced theories to explain all the facts so far given, and several others which I have not mentioned.

There is one set of facts, however, which seems to me conclusive. We can calculate the times of eclipses for some years ahead with an error seldom more than five seconds. We can also calculate the times of past eclipses. When we do so by the same methods we may be several hours out, when we get back to eclipses over two thousand years ago whose times have been recorded. This error is cleared up if we suppose that the earth's motion round its axis is slowing down, so that every hundred thousand years or so the day is a second longer. This slowing is exactly accounted for by the friction of the tides, which act as a brake on the earth. Of course the movement of the firmament might be slowing down. But it would be an altogether incredible coincidence if it were slowing down at just the rate calculated from the known facts about tidal currents and the known mass of the earth.

I will next deal with the evidence that the earth really does go round the sun, and that this is not just a convenient way of explaining astronomical observations.

2

The Earth goes round the Sun

I SHALL now try to give the evidence for the statement that the earth goes round the sun once a year in an orbit nearly two hundred million miles across. The evidence is not quite so strong as for the theory that the earth turns round once a day, because linear motion is not so easy to detect as spin.

If you look at the moon at the same hour on consecutive nights, you see that it lags relatively to the sun. Each day it rises and sets, on an average, nearly an hour later. Similarly the sun lags about four minutes a day relatively to the stars, and the other planets lag by different amounts. Naturally we set our ordinary clocks by the sun, not by the fixed stars. But the stars keep much better time than the sun. The length of a sidereal day, that is to say the time between two occasions when a "fixed" star is in the same direction relative to objects on the earth, is extremely steady. The length of a solar day varies throughout the year.

All these things are explained, and can be very accurately calculated, if the earth and planets move round the sun, and the moon round the earth, in elliptical orbits, according to Kepler's laws. There are two corrections to be made to these. The orbits are not exactly ellipses, because the planets are attracted by one another according to Newton's laws as well as by the sun. And Einstein's modification of Newton's laws embodies a much smaller correction. Still the apparent motions would be just the same, though no simple theory could explain why they occurred, if the earth were still, the sun moved round it, while the other planets moved round the sun.

There are however several facts which do not fit in with this view. One is the fact called aberration. If rain is falling vertically and you run through it, it seems to slope towards you. In other words it seems to be coming from in front of you.

Starlight behaves in the same way. The apparent directions of the fixed stars vary once a year. Those of all the distant ones vary similarly, and the variation, which is called aberration, is just what is calculable from the speed of light and the speed of the earth in its orbit. Aberration would be quite unintelligible if the earth were fixed.

Besides aberration, there is another apparent annual motion of the nearer fixed stars, which is called parallax. They seem to shift, in the course of a year, against the background of the farther ones. This confirms the theory of the earth's motion, but does not enable us to measure it, because it is only by measuring parallax that we can find out how far off the stars are. Still another light effect tells the same story. If we are moving towards a light, any particular line in its spectrum is more refracted, and would even appear bluer if the speed of approach was very great. This is because more waves reach us from it in the course of a second. Now the earth is moving towards those stars which we see to the south about 6 a.m., and away from those which we see to the south about 6 p.m. And the light from the same star is measurably, though not perceptibly, bluer when we are moving towards it than when we are moving away.

Yet another bit of evidence comes from shooting stars or meteorites. These are bits of stone or metal which fall into the air from space. We should expect to sweep up more of them in the part of our atmosphere which is moving forwards than in the part which is backing. And so we do. Not only are meteorites commoner just before dawn than just after sunset, but they are moving faster. It would take a very ingenious theorist to explain why this should be so if the sun moved round the earth.

The scale of the solar system can be measured in many different ways. It is quite easy to find out how far off the moon is. You can photograph it simultaneously against a background of stars in England and South Africa, and see at once that it has shifted, as a near object changes its position relative to its background if you look at it first with your right eye and then with your left. It is not so easy to find out how far the sun is. Occasionally the planet Venus passes directly between the earth and the sun, and can be seen as a black dot. By comparing the time taken by Venus in a transit, as this phenomenon is called, as seen from different parts of the earth, the sun's distance was fairly accurately measured in the eighteenth century. Since then it has been measured much more accurately by observations on several minor planets too small to be seen without a telescope, which sometimes come very much nearer to the earth than Venus ever does.

The aberration of starlight gives us yet another measure, not quite so accurate. So does the lag of about 16 minutes in the times at which Jupiter's satellites are seen to be eclipsed when the earth is farthest away, compared with when it is nearest. The important point is that all these methods tell the same story. The fact that they do so makes us trust the methods concerned when we can only use one at a time, as is the case when we get to the distances of the "fixed" stars.

There is one set of methods which tell us nothing. We get no answer if we try to find out how quick the earth is moving through space, though we can find out how quick it is moving to or from another body. This fact is the basis of the theory of relativity, which means, in simple words, that space is less real than matter. Strangely enough, however, though there does not seem to be such a thing as absolute position, motion or rest, there is such a thing as absolute direction. One can detect spin apart from any influence of external bodies.

Incidentally, the scale which we find for the solar system tells us that the earth is an average sort of planet. Mercury, Venus and Mars are smaller, Jupiter, Saturn, Uranus and Neptune bigger, Pluto about the same size. The mountains on the moon are about as high as those on earth, though much steeper, as there is no rain to wear them down. All this makes sense. The properties of matter seem to be just the same in one part of the universe as in another. Common sense sometimes lets us down, but not very often.

In the next few years it is probable that we shall know these distances much more accurately by radar. Echoes from the moon have already been picked up, and it should be quite possible to measure the distance of the nearest part of its surface within a few miles. A similar experiment with Mars, Venus or a minor planet would be vastly harder, but not necessarily impossible. The sun is already sending out so many radio waves that it would be much harder to pick up echoes from it with the necessary accuracy.   We shall next deal with distances and sizes of things outside the solar system, and see that they make sense too.

3

The Nearer Stars

THE nearest star to our solar system which is at present known is in the constellation called the Centaur, so far south that we never see it in England. This star is about 1¼ parsecs away. Astronomers measure distances in parsecs. It takes light about 3¼ years to travel a distance of a parsec.

At a distance of one parsec the position of a star relative to the more distant ones behind it, shifts through an angle of one second relative to the more distant ones behind it as the earth moves round the sun in the course of a year. This is one three thousand sixhundredth of a degree, or the apparent size of a halfpenny three and a quarter miles away. A shift of far less than this is readily detected on comparing photographs taken at intervals of six months. Most of the nearer stars also move quite perceptibly against their backgrounds in a few years. This effect is due to their own motion, not to the earth's. Thus if we know their distances we can calculate the speeds at which they are moving at right angles to the line of sight. These speeds are reckoned in miles per second, which compares with the earth's speed of 17 miles per second along its orbit.

Now we can also measure the speeds of these fixed stars along the line of sight by means of the Doppler effect, that is to say the blueing of light from an approaching object, and the reddening of light from a receding one. The average speed in the line of sight is a little less than the average speed at right angles, so the whole calculation makes sense.

Again we can calculate the amount of light which a star would give if we were as near to it as we are to the sun. One finds that some stars are much brighter than the sun, and some much dimmer. Now the stars can be classified both by their colour and by the elements which can be detected in their atmospheres with a spectroscope. Some of them have spectra very like the sun.   All the sun-like stars whose distances are known turn out to give out light at about the same rate as the sun.

In fact the results of astronomy go on making sense.   The sun is a fairly typical member of a large class of natural objects. Most results of scientific research of this kind.   They check up on one another, and make nature appear more natural.

But enough strange and unexpected results are known to keep scientists busy.   The same calculations which showed that the sun is a typical star of its kind showed that some other stars must be enormously bigger than the sun.   We know their surface temperatures, and therefore the amount of light which they emit per square mile, from their colours.   Some which have cooler surfaces than the sun, put out thousands of times more light per minute.   So they must be so large that there would be room for the earth's orbit inside them.

If so it should be possible to measure their diameters, not directly, but by the phenomenon called interference, which is also used for measuring very small distances, for example the thickness of a soap bubble.   It was used to measure the very small angle subtended at the earth by the giant red star Betelgeuze in Orion, and gave the expected result.   Since then several more diameters of large stars have been measured in the same way.

Something like a third of the stars are probably doubles, that is to say they consist of two sun-like objects moving round their common centre of gravity in ellipses, as they should according to Newton's laws.   If we assume that the relation between mass and gravitational force is the same among the distant stars as in the solar system, we can determine the masses of a pair of stars whose distance is known.

No one was surprised when it was found that stars with the same colour and spectrum as the sun had about the same mass as the sun.   But astronomers were very surprised indeed when they found that all the stars without exception had about the same mass as the sun.   The agreement is only rough.   Few stars are ten times as massive, none known to be a hundred times as massive, as the sun.*   Some have only about a tenth of its mass.   The

* At least one is now known with over 100 times the sun's mass.

luminosity of a star, however, increases much more than in proportion to its mass. A star ten times as massive as the sun gives out about 1,000 times as much light.

The reason for the close relation between mass and luminosity was largely explained by Eddington, but is a little too complicated to give here. The reason why no stars are much larger than the sun is a simple one. A very large star would generate so much heat that it would burst, and the double and triple stars are probably stars which have burst. So perhaps are clusters of stars like the Pleiades. Other stars are quite unexpectedly small and dense. There may be plenty of stars whose mass is less than a tenth of the sun's, but if so they give out so little light that we have not yet detected any of them.

Once the relation between mass and luminosity for the stars of known distance and mass was known, it was applied to the stars in general. In fact we have only to measure the brightness of a star and to observe the spectrum and colour, to tell roughly how large it is, and how far away.

However, before this relation was known, another method was available. Some stars are variable because a companion star occasionally eclipses them, others because they pulsate regularly, swelling up and collapsing again in the course of a few days. These are called Cepheid variables, after Delta Cephei, a star in the Milky Way not far from Cassiopeia. There are a lot of Cepheid variables in one of the clusters in the southern sky called the Magellanic clouds. Miss Leavitt found that all those in this cluster with the same period had the same brightness, and that brightness and period varied together in a simple way.

As the distances of some of the nearer ones were measured, it was found that this was a general rule. So one can calculate the distance of any Cepheid variable when one knows its period. As all Cepheid variables are very bright stars, this enables us to measure distances much too great for the parallax method. In fact they have been used to measure not only the sizes of the Milky Way but the distances of the nearer galaxies outside it.

The main difficulty with these indirect measures is that there is a good deal of dust in the space between the stars. When I say a good deal I am speaking in a "Pickwickian sense". The light from a star has to travel for thousands of years through one of these dust clouds before half of it is stopped. However the spaces concerned are so vast that the amount of matter in these dust clouds may be greater than that collected in stars. It is possible that this dust is constantly forming new stars by condensation.

All these discoveries about the distances of the stars hang together to tell a coherent and fairly simple story. But when we get to still greater distances there are real difficulties, with which I shall now deal.

4

Seeing the Past

THE stars in our immediate neighbourhood, including most of those which we can see, are about equally dense in all directions. But when we look through a telescope we see that the Milky Way consists of millions of stars, and that faint stars become commoner as we get near to it.

Besides this general concentration of stars there are star clusters of two types, namely open clusters like the Pleiades, and dense or globular clusters of which the few visible with the naked eye look like single stars, and only a telescope shows that they consist of many thousands of stars. These clusters contain Cepheid variables which enable us to measure their distances in the way which I explained in the last article. The open clusters are near the Milky Way, and the farthest yet detected is about 500 parsecs away. In other words the light which we see from it started out about A.D. 300. The globular clusters are much further off, up to 50,000 parsecs, and are not found near the Milky Way. The light from them always started before the beginning of human history, and generally during the Ice Ages.

A careful statistical study of star motions makes it very probable that the stars in our neighbourhood, including the sun, are moving round the centre of the Milky Way, which is in the constellation Sagittarius, probably in spiral rather than elliptical orbits. The distance of the centre is about ten thousand parsecs, and we go round it once in two or three hundred million years. That is to say we have been round about once since the Coal Measures were formed.

The whole system of all the stars which we can see with our eyes or with an ordinary telescope is a biscuit shaped object about five thousand parsecs thick, and perhaps thirty thousand across. The sun is about half way between the centre and the edge. The Milky Way is just the appearance of immense numbers of stars which we see when we look in the plane of the "biscuit". The globular clusters lie out in a sphere roughly enclosing the biscuit. The mass of the whole system is something like two hundred thousand million times that of the sun, and there are probably about five times that number of stars in it. This is about a thousand times the total number of men and women, and about equal to the total number of birds.

In the direction of the Milky Way we cannot see what is beyond it, owing to the clouds of stars and of dust. But in other directions we can see a few rather faint objects, and photograph tens of thousands of them. These are the so-called spiral nebulae. The most easily seen is in the constellation Andromeda, and though it is very faint, its apparent size is larger than the moon's. With a good telescope it is seen to consist of stars arranged in a rather irregular spiral. Among the brighter stars in it are some Cepheid variables, so its distance can be measured, and consequently its actual size. It turns out to be about as large as the Milky Way, and to be so far off that the light which we now see from it started about nine hundred thousand years ago, that is to say at the end of the Pliocene era, before the Ice Ages, and before we know of any fossils or tools which are certainly human.

The distances of hundreds of other spiral nebulae have been measured with less accuracy. They are all of about the same size. When their spectra are photographed, they are all seen to consist of the same sorts of elements as are found on earth, and to be spinning round like so many catherine wheels. They are quite unlike the stars in one important respect. The distances between stars, even in a dense cluster, are very much bigger than their diameters. On the other hand the distances between neighbouring spiral nebulae are often only about twenty times their diameters, and sometimes less.

There is, however, one very queer thing about them. The spectroscope shows that the light from them is reddened, not by a scattering of the blue light, as in the case of the setting sun, but by lowering of the frequencies of vibration of light of all kinds. This could be explained if they were all moving away from one another, that is to say if the universe were expanding.

If so the farthest ones yet photographed are moving away from us at about an eighth of the speed of light.

It could also be explained if in the course of time the movements of electrons inside atoms were speeding up, so that a particular type of event in an atom, for example the approach of an electron towards the nucleus of a sodium atom which gives the yellow light of some street lamps, gave out more energy to-day than in past geological ages. Milne showed that these two ways of describing the same events are not really different, but depend on different ways of measuring time and space. In fact an extension of the theory of relativity seems to solve this problem, though it raises new ones.

The farthest spirals so far photographed are so far off that the light which reaches us from them has been travelling since Permian times, just after the coal was formed. With the great new mirror which has been made for the observatory at Palomar in California, it will be possible to photograph light about twice as old, that is to say Cambrian light. It may well go still further into space, and back into the past.

This will allow us to measure the rough distances of most of the nearest million or so spirals. According to some theories, for example Eddington's, the total amount of matter in the universe is finite, and if we could photograph things only four times as far away as this we should find they were moving away with almost the speed of light. According to other theories, such as Milne's, the amount of matter in the universe is infinite. This of course could never be proved. But it may be possible, within our lifetimes, either to prove that Eddington was wrong, or that so far his theory has led to correct predictions.

The important point is that such questions as these are not questions to be argued about forever by philosophers, but questions to be settled by observation. The apparatus needed will cost some millions of pounds; but the settling of such questions has always given us enough new knowledge about matter to be well worth while.

In science we always find regions where our knowledge is clear, and others where it is far from certain. One of these is of course the study of things very far off in space or very far back

H

in time.   These are really the same problem in different words. When we can photograph objects eight or ten times as far off as now we shall be photographing them as they were at about the time when the earth was formed.   Another region of uncertainty is concerned with extremely small distances and times.   We get into this region when we study the nucleus of an atom.

There is still a third region of uncertainty.   We can see things a little smaller than a wave length of light with a microscope, and we know about atoms and molecules from chemical experiments.   But it is harder to find out about things larger than molecules but much smaller than wave lengths of light. And it is just in this region that the transition between chemistry and life occurs.

My own work lies largely in this third region.   But it is amazing how rapidly our knowledge has extended, both concerning large and small things, in my own lifetime.   We have got somehow to make this knowledge part of the heritage of ordinary people.   The best way to do this is not to use large numbers, but ideas which should be familiar to every educated man or woman, such as the size of the earth, or the time since the coal or chalk were formed.

V

ASTRONOMICAL ESSAYS

I

Common Sense about the Planets

IN the development of science there is a constant and fruitful struggle going on between two tendencies. On the one hand we should like to explain everything, that is to say to find a reason for it. Geologists do not believe that it is either because of pure chance, whatever that may be, or an arbitrary act of creation, that the Thames bends to the right between Lambeth and London city. But they would equally agree that we do not yet know why it does so, and that any theory is likely to be wrong, and if generally accepted, likely to hold up the progress of geology. So although it is an excellent thing to find the right explanation, it is quite reasonable to say that we do not know the explanation, and had better for the present confine ourselves to describing things as they are.

A very good example of this fruitful strife is found in the history of our ideas about the solar system. For thousands of years men have known that they could see the sun, the moon, and five other planets. Actually you can sometimes see an eighth, Uranus, if you know where to look. Early men said there were seven planets because there were seven gods each identified with one of them. When their motions were described people asked why they moved as they did, and why there were just so many.

Kepler tried to answer the question. He thought that Mercury Venus, Earth, Mars, Jupiter and Saturn each moved in a sphere with the sun as its centre, and the spheres were so arranged that one of the regular solids would fit in between two of the spheres. As there are only five kinds of regular solid, of which the cube is the best known, there could only be six planets. Kepler found that this theory wouldn't work, and later found the correct shape of the orbits, or very nearly so.

Newton showed why the orbits have the shape they have, and

why a planet at a particular distance from the sun must move at the speed it does.   But he did not try to explain why the known planets were at the distances calculated.   He thought that this was due to the act of the Creator, though once created, their later motions could be explained.

Hegel explained why there were just seven planets.   Unfortunately for him, a lot more have been found since, and no one takes his theory seriously.   But he was quite progressive in supposing that the structure, as well as the motions, of the solar system, must have a rational explanation.   An astronomer called Bode produced a "law" or rule for the distances of the planets from the sun.   But it is only very roughly true, and if it were quite true, no one has suggested a reason why it should be.

This year a bit of real progress has been made by Kuiper, of Yerkes Observatory, who points out that there is a relation between the masses of the planets and their distances, and that this also holds for the satellites of the great planets Jupiter, Saturn, and Uranus.   If the planets were formed from the condensation of dust or gas whirling round the sun, or from the breaking up of a filament shot out of the sun when another star passed near it, or in several other suggested ways, then the heavier a planet, the bigger the gap should be between it and its neighbours. For a heavy planet would not only use up more of the available matter, it would attract the matter in its neighbourhood, and prevent the formation of other planets near it.   There are other more complicated arguments which suggest what the relation between masses and gaps should be.

So here is what Kuiper did.   He added together the weights of each pair of neighbouring planets, for example Venus and Earth, Earth and Mars, and divided by the weight of the sun.   He also calculated the differences between the average distances of these planets from the sun, divided by the sums of these distances. So he got two columns of figures, and found that they fitted pretty well to a relationship derived from theory.   Then he did the same for the systems of moons moving round Jupiter, Saturn and Uranus.   He found that the numbers followed the same rule.

In fact the rule which they fit is not quite what would be

expected on any theory yet suggested, though one simple theory is nearly right.   The rule finally arrived at is not completely accurate either.   In fact only about half the distances calculated are within ten per cent. of the true values.   But that is something to be going on with.   Fortunately for science, we shall soon have the means of finding out whether Kuiper's rule is really universally valid.   Slight irregularities in the motions of some of the so called fixed stars, which of course are distant bodies like our own sun, seem to show that they too have planets.   When the distances and masses of these planets are measured, if they conform to Kuiper's rule we shall have to take it very seriously.   Meanwhile it may perhaps be made more accurate, by taking account of all the planets or satellites in a system, not only immediate neighbours, and by finding a more complete theory to account for it.

The important point is that Kuiper's work justifies the view held by all Marxists, and a great many scientists who are not Marxists, that there are no arbitrary features in the universe, that is to say facts for which no rational explanation can be found.   Of course at the present time we can only find a rational explanation for a tiny fraction of the known facts.   And no doubt a lot of the explanations which we give will turn out to be wrong when we know more facts and think more clearly about the known ones.   But we are right to ask why the earth goes round the sun once a year, and Jupiter once in eleven years, why male and female lions differ more in hair length than male and female tigers or cats, why sugar and glycerine taste sweet and alcohol does not, and so forth.   We don't know the answers to these latter questions, but such work as Kuiper's makes us hope more firmly that our descendants will know them.

Looking for Blue Stars

TWO American astronomers, Humason and Zwicky, of Mount Wilson Observatory, in California, have recently reported a hunt for faint blue stars in two areas of the heavens. One was in the neighbourhood of the Hyades in the constellation of the Bull, where stars are very thick. The other was in the constellation Coma Berenices, as far away as possible from the Milky Way. They photographed the same area of sky, first through a yellow glass and then through a violet one, and picked out the forty-eight stars which appeared much brighter through the violet than through the red screen. All these stars were very faint indeed. Even the brightest was twenty times too far away to be visible with the naked eye.

Readers may well ask what can be the point of such a search, and even if it has more point than looking for the bluest postage stamps what possible practical value it can have. The answer is quite simple. The bluer a star, the hotter it is as a general rule. The reason is straightforward. As one heats a black object up it begins to produce invisible infra-red rays which can be felt as heat. Then it becomes red-hot, and later white-hot. Still hotter things give blue light. Thus an arc lamp is much hotter than a filament lamp, and much bluer. Of course many modern lamps, for example neon, mercury and sodium lights, shine because the atoms in them are electrically excited, not because they are hot, and their colour obeys quite different rules. But since blue light consists of more rapid oscillations than red, it is quite natural that a hot body, whose atoms are moving quickly, should give bluer light than a moderately hot one.

So what Humason and Zwicky were really doing was looking for the hottest bodies in two regions of the universe. This is a reasonable thing to do, because matter under extreme conditions develops new properties (the change of quantity into quality).

For example some very cold metals become super-conductors with practically no resistance at all, so that an electric current induced in a lead ring in a bath of liquid helium will go on circulating for hours on end.

The spectra of these stars were photographed, which made a calculation of their temperatures possible according to a principle first laid down by the Indian astronomer Saha, and worked on in more detail by Milne and Fowler in England. In a number of cases their so-called proper motions, that is to say the rate at which their directions in space alter, were already known. This made it possible to pick out those which were near to the sun, near that is to say in an astronomical sense. For a near object does not have to be moving very fast to alter its apparent position noticeably in ten years or so. So most stars which change their position quickly are fairly near to us.

In fact most of the blue stars in the Hyades shared the motion of Aldebaran and the other bright stars of this cluster. So they must be very much dimmer. It turns out that they belong to the group of stars called White Dwarfs, which are extraordinarily dense.

The matter in them is so closely packed that a cubic inch of it would weigh a ton. It is of course held together by its own gravitation, which is so enormous that even light has a good deal of work to do in getting out of such stars, and gets redder in the process, though not enough so to stop the stars being bluish white. Although no such star is bright enough to be seen without a telescope, they are so common near the earth that it is quite possible that there are more of them than of all the other stars put together. Searches in regions like the Hyades where there are a lot of stars will help to decide whether this is so. If so they probably represent a late stage of the evolution of stars, and perhaps our sun will finally contract into one.

The search for blue stars near the pole of the Milky Way gave quite a different result. A few are dwarfs, but most are very bright and large stars at an enormous distance. In the direction of the Milky Way there is so much dust between the stars that one can usually only photograph stars, however bright, a few thousand light years away. At right angles to the Milky

Way one can see out into space beyond it.   And some of these large blue stars may be a good way out in the almost empty space between our galaxy and its neighbours which we see as spiral nebulae.

So when more is discovered about them they may help us to fix the scale of the universe more closely, as well as telling us more about very hot matter.   When I say very hot, I do not mean as hot as the matter in the first millionth of a second of an atom bomb explosion, or the matter of an exploding star, but the hottest matter which is on permanent display.

I don't know what will come out of such studies, or whether anything will.   I know that the study of matter in the sun told us of the existence of one new kind of matter, namely the gas helium, and taught us much about the behaviour of the commoner kinds.   In particular, in the sun and other stars we study matter in a much simpler state than is common on earth, that is to say as gas consisting of single atoms, many of which have lost one or more electrons.   Their study has been of immense help in laying the foundations of chemistry.

Whether knowledge obtained in this way will be used for good or evil does not depend on the astronomers.   Unfortunately so, because they are very internationally minded, much more so than chemists, for example.   Star mapping, and particularly eclipse observation, are international matters, planned for many years ahead by the workers concerned, and workers in other jobs might well learn a lesson from the International Astronomical Union.   In fact when miners, transport workers, or above all, agricultural workers, have as good international tie-ups as astronomers, our planet will be a much happier place.

3

· Radio from the Sun

AS radio signalling and broadcasting developed, attention was more and more directed to noise, that is to say unwanted disturbances. Some were shown to come from electric motors and dynamos, others from thunderstorms.

But finally it became clear that others came from outside our atmosphere. Jannsky's discovery of this fact was not at first followed up. But the development of radar and short wave transmission during the war led to a fairly intensive study. As it also led to secrecy the actual advance in knowledge was probably no greater than what would have been made in peace time. Now however a good deal of work is being done on the question, and a young and vigorous group of Australian physicists is playing a leading part in it.

Our atmosphere is opaque to most kinds of radiation. The sun sends out a lot of hard ultra-violet rays, and also very probably some X-rays. If we tried to live on the moon, which has no air, they would be very dangerous. Only two sets of rays get through in any quantity. One is in the visible region and the neighbouring parts of the infra-red and ultra-violet. We can see many of them, and photograph the others.

The other group consists of radio waves with lengths round a metre ranging from about a centimetre to ten metres or so. Some of these come from the sun, others from regions of the Milky Way where the stars are very dense. It is striking that whereas starlight is a negligible fraction of sunlight, the stars do not contribute much less than the sun to radio noise at normal times.

The sun's radio emission is of two kinds. One kind, with a maximum intensity about 1 metre, is fairly steady. The other kind, with a maximum about 5 metres, is very much stronger, but only occurs when there are large spots on the sun's disc.

If the same thing happened with regard to light, the sun would usually give white light, but from time to time there would be a terrific red glow.

On January 29th 1948, the Royal Society held a meeting at which Ryle and Vonberg, of Cambridge, communicated their results on accurate measurements, while Martyn, an Australian, produced a mathematical theory of the origin of the waves. Martyn's theory is something like this. The radio which we pick up in the absence of sunspots does not come from the luminous surface of the sun, but from the very thin atmosphere round it, which is called the corona, and can be seen when the sun is totally eclipsed.

When we go up in an aeroplane the air gets colder, but at great heights the air becomes very hot. The hot layers are the ionospheres, which reflect long radio waves downwards. The corona is the sun's ionosphere, and to judge from its light spectrum is enormously hot, with a temperature round a million degrees, and very strongly ionized. The radio emission is due to disturbances in it. Its electrical charge is so high, and the speed of its movements so great, that it gives rise to radio emission.

The difficulty in proving this, or any other theory, is that it is very hard to get exact directional radio reception. You are doing well if you can distinguish between sources a degree apart. If your eye were no better than this, you could not see a halfpenny at a distance of five feet. When the emission is pretty weak, as it generally is from the sun, things become particularly difficult, and Ryle and Vonberg worked out a most ingenious method for distinguishing solar radio from other noise.

If Martyn is correct, when we get a real radio telescope which distinguishes directions as well as, say a very short-sighted man can do, we shall "see" the sun as a fairly bright disc surrounded by a very bright ring. For most of the radio seems to come from the edges of the corona. This is borne out by the few observations made during eclipses. There is a partial radio eclipse before any of the sun's light is cut off, and the radio emission is never totally eclipsed. In future radio observations during total eclipses are likely to teach us as much as, or more than, observations made with the eye or telescope. Whether the emission caused by sunspots comes from the spot itself or the corona above it remains to be seen, As there will be a maximum frequency of sunspots next year, we ought to know fairly soon.

What impresses me most about this work is its philosophical importance. For thousands of years people only knew about the sun, moon and stars by their light and heat. They formed all kinds of fantastic theories about the heavens, some of which were incorporated in religions.

Then Newton showed that the tides were due to the gravitational attraction of the moon and sun. Later Pallen gave the first accurate account of meteorites, objects from outside the air. Thousands have since been examined, and found to be made of ordinary matter.

Now we have a fourth way of exploring the universe. I do not know what people will discover with it. If I wanted to be sensational I would suggest that the solar radio noise was the confused records of radio messages by intelligent beings—angels if you like—living in the sun. I think this is most unlikely. But I also think that when we have real radio telescopes we shall discover much queerer things than that with them.

The world is full of queer things, but they are not a bit like what our ancestors imagined. For example some viruses, such as that of lethargic encephalitis, can make good children into bad ones, as devils were supposed to do. But devils were pictured as like men, though excelling them in power and wickedness. Whereas viruses are half-way between living beings and chemical compounds, in fact much more alien to our ordinary thought than devils.

So I guess that astronomical radio research may, in time, tell us altogether fresh and unexpected facts about the universe.

4

Cosmic Rays

ON few subjects in modern physics is more nonsense written than on cosmic rays. To begin with, they are not rays in the ordinary sense of that word. They are a beautiful example of what you discover if you study the exceptions to ordinary rules. The whole electrical industry is based on following up two exceptions. Most unsupported bodies fall to the ground. But iron filings near a magnet, or scraps of paper near an electrified rod of sealing wax fall up.

When the usual rules governing the behaviour of electricity were worked out, another crop of exceptions turned up. Air ought to be a perfect insulator, and it may be for minutes at a time. But an air gap always starts leaking in the long run. Some of the leakage is due to ionization of the air by high-speed particles flung out by radioactive atoms. But it was gradually shown that most of it is caused by something coming downwards. Hess and Kolhörster found that the leakage increased tenfold when they went up in a balloon to the height of six miles. Actually we now know that these "rays" consist almost entirely of very rapidly moving particles, and that at sea level about six of them pass through a horizontal square inch every second.

Three different instruments have been used to detect them. The Geiger-Müller counter, designed for work on radioactivity, is an electrically charged wire in the centre of a metal cylinder in a glass tube which discharges whenever a particle passing through the tube makes the air conduct electricity, or in technical language, ionizes it.

The second is the cloud chamber invented by Wilson. A vessel with a glass window contains air saturated with water or some other vapour. It is suddenly cooled by expansion, and the vapour condenses on any dust particles which may be present, forming a fog. After a few expansions all the dust is got rid of, and the condensation occurs on the trails of ions left by rapidly moving particles.   These tracks can be photographed, and as soon as Skobelzyn, who recently represented the Soviet Union on the atomic energy committee, did so, he found that the so-called rays were really particles.   Quite recently Powell, of Bristol, has developed special photographic plates in which cosmic rays make tracks which can be examined with a microscope.

The story of the analysis is told in Janossy's recent book.* Readers who know a little physics will find it good reading, but unless you know that MeV means a million electron-volts you are liable to stick on page 2, and there are several mistakes, notably the definition of a gamma ray on page 5.

One of the first steps was to couple up a number of Geiger-Müller counters in series, so that a current only flows if they all start to leak at once.   So if we have four such counters in a row it is very unlikely that current will leak in all four at once unless the same particle has gone through them all.   Move one out of the row, and such coincidences become much fewer.   If you put a screen of a couple of inches of lead above the counters, there are fewer discharges.   If you put it below them, nothing happens.   This shows that the particles come from the sky, not the earth.   Very elaborate batteries, sometimes of dozens of these counters, have been set up, and it has been shown, for example, that when particles are stopped by a screen of lead, they often generate a shower of other particles, which may set off as many as five counters on the same level at once.

The modern theory of cosmic rays is very largely due to Blackett of Manchester, and a group of colleagues, many of whom, like Rossi and Occhialini, were refugees from fascism. Blackett and Occhialini first coupled up a Wilson cloud chamber with Geiger-Müller counters so that a photograph was taken just when a particle had passed through a particular set of counters.

In this way they could pick out, say, those particles which could just get through an inch of lead, but not two inches. Further, they could find out how fast the particles are going by various methods.   The easiest to understand is perhaps the use

* *Cosmic Rays and Nucliai Physics*, by L. Janossy, Ph. D.   Pilot Press Ltd., 9/6.

of a magnet.   A magnet will pull an electrically charged particle out of a straight path, but the quicker the particle is going, and the heavier it is, the less the pull.

The world's most powerful permanent magnet used to be in Blackett's laboratory at Birckbeck College, London.   Now the most powerful ones are on Mount Alaghez in Armenia, where Alikhanyan and Alikhanov are working.   Both are Armenians, and they are brothers, but one of them put a Russian ending to his name to avoid confusion.

When cloud chamber photographs were examined, the surprising result emerged that the "rays" consisted of several quite different kinds of particles.   Some of these were electrons with positive and negative charges.   But there were also much heavier particles.   It was first thought that they were all of one weight, and they were called mesons.   Now it is generally admitted that there are two different sorts of meson.

However Alikhanyan and Alikhanov say that there are at least fifteen different kinds of particle heavier than electrons, some being as heavy as ordinary oxygen or nitrogen atoms. They are moving at fantastic speeds, very near to that of light, and break up in less than a millionth of a second. But in that time they can go through hundreds of feet even of rock.   Unfortunately the Iron Curtain seems to operate at Dublin, where Janossy works,* for he does not refer to this recent Soviet research, though it was published before he wrote his book, and I should be the last to accuse him of suppressing it.

These particles, heavier than electrons, but most of them lighter than ordinary atoms, play a part in building up atomic nuclei, and seem to be among the things emitted when they explode.   In fact their discovery links up with the research which has so far only given us atomic bombs, but will give us power for peaceful purposes also.

The origin of cosmic rays is completely mysterious.   The original particles which cause them are mostly stopped high up in the air, and the particles observed at sea level are due to their hitting atoms on their way through the air.   They do not come equally from all directions.   The earth's magnetic field

* He has since gone back to his native Hungary.

deflects some of them.   But when this is allowed for they do not
come from anywhere particular, for example from the sun or the
Milky Way.

This fact is a serious gap in our knowledge of the universe,
seeing that the energy absorbed by the earth from them is rather
more than it gets from the light of all the stars together.   In
fact we are only at the beginning of our knowledge of them.
We no more know where it will lead than we knew in the case
of radioactivity fifty years ago.   It is up to us to see that we make
the kind of world where this knowledge will be used for human
happiness and not for destruction.

5

Why the Earth is a Magnet

DURING the nineteenth century physicists were constantly discovering new facts linking together the different properties of matter. Thus electric currents were found to produce heat and magnetism according to definite laws, and conversely temperature differences and changing magnetic fields were found to generate electric currents. These facts were discovered by such men as Ohm and Faraday as the result of dramatic experiments which could readily be repeated. These experiments gave the basis for much of modern industry; for example dynamos and electric stoves are merely developments of apparatus first set up to demonstrate the connection between different properties of matter.

In the twentieth century similar discoveries are made. But they can rarely if ever be demonstrated by a simple laboratory experiment like the glowing of a wire or the movement of a compass needle when a current is switched on. This is because they refer to the properties of very small or very large objects which only give effects on an ordinary scale indirectly. Sometimes this is because they are very small. For example the hot filament in a radio valve shoots out electrons which carry a current that can be varied by changing the potential of another part of the valve. But to understand what is going on, one has to learn to think in terms of electrons, which are much less familiar than electric currents, and much harder to demonstrate to the human senses. Sometimes the objects concerned are very large. Thus it has been shown that gravitational fields can bend light, and change its colour. But this was only proved by observations on the sun and stars using telescopes, and photographic plates on which careful measurements were made.

On May 15th 1947, Professor Blackett announced to the Royal Society what may be a new connection between the properties of matter, namely that all spinning bodies produce a magnetic field. He did not produce any experiments to prove it, but based his argument on the magnetism of the earth, the sun, and certain stars. It has of course long been known that the earth is a magnet. That is why one can find one's way with a compass.

The sun was first suspected to be a magnet because the streamers in the corona, which is seen round it during an eclipse, have a pattern like that which iron filings take up in the neighbourhood of a magnetized steel ball. It was proved to be a magnet after Zeeman had found that the spectrum of an element is altered by a magnetic field. The lines in the solar spectrum are altered in the same kind of way as those in a gas flame coloured by sodium if it is held between the poles of a magnet. The sun is a much more powerful magnet than the earth. In fact its magnetic field is about a hundred times as strong, and its magnetic moment is a hundred million times that of the earth.

This year Babcock of Mount Wilson Observatory, America, examined the spectrum of a star which on other evidence was believed to be rotating rapidly, and found the same distortion of the spectral lines, but enormously stronger than in the sun. The star in question turns round its axis about once a day, and the magnetic field appears to be about 1500 gauss. This is a considerable field strength. If the earth had such a field, steel ships would set themselves north and south like compass needles, and it would be very hard to steer them.

There is evidence that some small stars have magnetic fields far stronger than this, in fact stronger than any which have been made in a laboratory, let alone used in industry; but they have not yet been measured. The measured magnetic fields cover a range of 2,500 fold. The magnetic moment of Babcock's star is about ten thousand million times that of the earth.

Blackett's discovery is simply that the magnetic moment is exactly proportional to the angular momentum. That is to say if we know the mass and shape of a body, and the rate at which it is spinning, we can calculate its magnetic moment. To be quite accurate, the magnetic moment of the star is forty per cent. more than it should be if calculated on the basis of the earth's, but the measurements are not yet very accurate, and an error of forty per cent. in a factor of ten thousand million is not very serious. In other words every rotating body is a magnet. The magnetism cannot be due to rotating electric charges. If the earth had the charge needed to make it a magnet for this reason, there would be an electric field of millions of volts per inch at the surface.

If this theory is true, it may be asked, why can it not be demonstrated in the laboratory? The test would have to be made with a fly-wheel or sphere spinning so fast as to be very near bursting. Steel could not be used, and a non-conductor would probably have to be tried as well as a metal. Finally the field produced even by a bronze sphere ten yards in diameter spinning as fast as it safely could would only be about a millionth of a gauss, and extremely difficult to detect at all, let alone to measure accurately. It is unlikely that the experiment will be made for many years.

The theory may facilitate work on atomic nuclei, by helping us to understand how a neutron, though it has no electric charge, behaves as a magnet. More likely it will help in the framing of a cosmology which will link together gravitation and electromagnetism more satisfactorily than Einstein's theory of general relativity.

It is of some interest that Blackett carried out this work while president of the Association of Scientific Workers. The sort of man who becomes president of this union is one who does not keep his eyes glued to a particular job of work, but looks around him. So is the sort of man who finds out new facts about the universe.

VI
AN OUTLINE OF ZOOLOGY

I

The Classification and Origin of Animals

I AM constantly being asked questions about animals, and it is fairly clear that my questioners have never learned the elements of zoology. That is a severe criticism of our educational system. There are two good reasons why we should know about animals. The first is that many of us have to deal with them, though in modern cities most people have rather small chances of doing so. The second reason is that we are animals, though animals of a very peculiar sort, and we cannot understand ourselves and one another without understanding what we have in common with other animals.

One of the things which zoologists have to do is to classify animals. Every animal is assigned to a species. Roughly speaking a species means a group of animals which will mate without difficulty and give fertile offspring. Each species is given two Latin names. For example the sewer rat is called *Mus norvegicus*. The first name is the name of the genus *Mus*, which also includes the ship rat *Mus rattus* and the house mouse *Mus musculus*. They may be compared to human names like John Smith and Robert Smith. However a species is, or should be, a biological reality, a genus is a matter of convenience. Some authors put the house mouse in a different genus to the two rats.

Similar genera are put in the same family, similar families in the same order, similar orders in the same class, and similar classes in the same phylum. For example the genus *Mus* is assigned to the family *Muridae*, including various rat-like animals. The family is included, with squirrels, porcupines, beavers, and so on, in the Order *Rodentia*, or gnawers. The rodents are included in the class *Mammalia*, hairy warm-blooded animals which suckle their young. The *Mammalia* are one class of the phylum *Vertebrata*, which agree in having backbones, and in many other characters. There are also intermediate divisions, such as suborders and subclasses.

Now when such classifications were first made they were

made for convenience, like the classification of words into nouns, adjectives, and so on. But when the theory of evolution was accepted, zoologists tried to classify together animals which they believed had a common ancestor. For example it is probable that all mammals are descended from members of one species of reptile whose organs underwent certain changes, while others stayed put. This is quite different from the classification of words, chemical substances, or human beings. Nobody thinks that all adjectives are descendants of a single adjective. On the contrary adjectives like wooden are made from nouns, and adjectives like eatable from verbs. We do find a trace of this idea when for example the Jews are described as the children of Israel, although we know that plenty of people not descended from Israel became Jews, so that Jews are of mixed origin.

However the classification of animals is not as logical as it should be and probably will be when we know more. For example the class of Reptiles has given rise to Mammals and Birds, as we know because we have found fossils of intermediate forms. In fact a bird is more nearly related to a crocodile than a crocodile is to a tortoise. That is to say you have to go farther back into the past to find the ancestor of all three groups than the common ancestor of birds and crocodiles. So the class of Reptiles is a group of animals which have stayed at a certain level of organisation, rather than all the living descendants of a common ancestor. A group made up in this kind of way is sometimes called a Grade.

In this series of articles I am going to describe the main groups of animals. But before I do so we must ask what an animal is. The simplest answer would be that an animal is a living thing which can move itself about. This is generally true. But most animals pass through a stage where they cannot move. You and I could not move at all in the first few months of our lives before birth, and not very much in the later ones. An oyster swims about vigorously when quite young, but passes most of its life stuck to a rock. More serious still, some microscopic plants swim about throughout their lives, and some quite large plants, such as seaweeds, start life as microscopic swimmers.

Others have tried to make the distinction on the ground of nutrition. Most plants use sunlight and carbon dioxide to make their living substance, while all animals have to feed on plants or on other animals. But so do the fungi and moulds, which are classed as plants.

In fact no sharp distinction can be made. There is a group of very small living beings called the Flagellata. They are single cells, often shaped rather like a tadpole, and swim about by lashing their flagellum, which corresponds to the tadpole's tail. Some of them are certainly plants. They only need water, light, carbon dioxide, and salts. Others are certainly animals. They have mouths, and eat other plants and animals. Some live on food in the water, but have not got definite mouths.

Biologists mostly believe that the animal forms have evolved from the plants. For plants must have existed before animals. Plants can live without animals to eat them, and animals need plants as a source of food. Quite recently it has become possible, in this group, to make animals out of plants in the laboratory. The attempt very often fails. Some plant flagellates, of the kind which make the water in ponds go green, stay green when kept in the dark and fed on sugar. Even after several years, and thousands of generations, in the dark, they stay green. Others lose their green colour, but regain it within a few hours when put back in the light. This is quite in accordance with Weismann's theories.

But when other kinds are kept in the dark, the number of green plastids which are the organs for using light to make sugar, diminishes. It may diminish so much that when a cell divides one of the daughter cells does not get a plastid. Here an organ has been lost through disuse, as Lamarck and Darwin taught that organs could be lost. The descendants of such a cell never become green, and starve to death even in light, unless they can get food.

It is probable that animals originated from plants on a great many occasions, generally by a sudden "leap" of this kind, and it is likely that the first animals were flagellates. But it is fairly sure that, even if animals arose separately several different times, the first animals in each pedigree consisted of single cells. The single celled animals, and others consisting of a number of cells not much differentiated, are called protozoa, or First Animals. Most of them are too small to see without a microscope. But they can build mountains and overthrow empires, so they are quite important.

2

The Simplest Animals

THE simplest animals, which are called the protozoa, consist of single cells. There are several good reasons for thinking that the first animals did so. There is no sharp gap between the single-celled animals and plants living to-day, and a very big gap between the many celled ones. Every animal except a few which reproduce by splitting or budding, starts life as a single cell. You and I did so. Since the development of an individual runs roughly parallel to the evolution of its ancestors (for example you and I had tails and hairy coats before our birth) this is an argument for a one-celled ancestor. And fossils of protozoa are found in pre-Cambrian rocks before anything but fragmentary fossils of other groups are found. They are certainly as ancient as many-celled animals, and probably more so.

The protozoa living to-day are grouped into four classes. Perhaps the most primitive class is the Flagellata, which swim about by lashing their tails. You will find them in all sorts of foul water. But you may also find them in human and animal blood. A group of them, called the trypanosomes, are one of the curses of Africa. They cause "sleeping sickness" in man, and a number of diseases in animals.

Many people regard the class called Rhizopoda as even more primitive, though others think their ancestors were flagellates, for some of them have flagella when young. They include animals such as *Amoeba* which creeps about in damp soil. It is a mass of protoplasm constantly changing its shape, and lives by engulfing smaller creatures, such as bacteria, or fragments of larger ones. *Amoeba* is too small to see clearly without a microscope, but some other rhizopods are quite large, and build complicated skeletons of limestone or silica. The largest were the nummulites which lived in the sea some fifty million years ago. They ran up to about the shape and size of a penny, and their skeletons have formed ranges of limestone hills. The third class are called the Sporozoa, which are all parasitic, and reproduce by very small spores. Compared with Sporozoa, tigers, wolves, and crocodiles are pleasant and harmless animals. One group of them cause the set of diseases known as malaria in man. They live in the blood, and every one, two, or three days, they burst out of the red cells where they have been growing, and swarm. This causes a violent bout of fever. I have no doubt that malaria made history on a very large scale. I don't believe that British are a superior race to Indians. But I do believe that a Briton without malaria is generally superior, particularly as a fighter, to an Indian infected with it.

Malaria is terribly common in most warm countries where mosquitoes can live. For the sporozoa concerned live in mosquitoes as well as man, and when a mosquito bites a man they can be transferred from one to the other. Socially malaria is a worse disease than plague or small-pox, because it lasts for years, and may turn most members of a population into chronic invalids. Those parts of India which are chronically infected with malaria have frequently been conquered by quite small armies from less malarious countries such as Afghanistan, Nepal, France and England. One of the main tasks of an Indian government will be to rid their country of this disease. This will be a much more efficient measure of national defence than making aerodromes or warships. Other sporozoa attack animals, and are responsible, among other things, for some of the worst diseases of rabbits and poultry.

The fourth class is called the Ciliata. They are mostly swimmers, their bodies covered with tiny structures called cilia, which beat in unison like oars projecting from a boat. They have very definite shapes, and parts of their cells are specialized to perform functions like those of the mouths, stomachs, kidneys, and so on, of higher animals. They may even have the beginnings of a nervous system. Most of them live in dirty water, and some can just be seen with the naked eye as tiny dots. Their shapes are perhaps the most fantastic of all those which animals can take.

For one thing they are often quite asymmetrical. This is

an advantage to a swimming animal which has no eyes or ears to guide it, though it has something like a sense of smell or taste which attracts it to the neighbourhood of food. A symmetrical animal swimming blind will go in a straight line or a circle, so it will either leave the favourable area quickly, or come back to where it started. An asymmetrical one moves in a corkscrew spiral and can visit most parts of a small volume of water in the course of time. Some of the ciliates spend most of their lives attached to a plant or stone, and catch their prey as it goes by. Only a few live in larger animals, and cause disease.

The protozoa usually reproduce in a very simple way. A cell eats till it has doubled its size, and then splits in two, and the new ones start afresh. Even kinds large enough to see can double their size in a day. But sometimes they have sexual reproduction, in which case two cells often fuse into one, which later divides. There is every gradation between complete union and a process like the sexual reproduction of larger animals.

We are just beginning to find out about their sexes. Only a few fall into two groups corresponding to males and females. But in some species there are several different mating types. Members of the same mating type never mate. But they can mate with members of any other type. If there happen to be just two mating types in a species we may call one males and the other females. But some species have three mating types, and others up to eight. None of the higher animals have more than two sexes. It is hard to imagine what human life would be like if we had more than two. Certainly there would be new plots for novels. For if three protozoa are of different sexes, any two can mate. But as no protozoan mates more than once there is bound to be an odd one left out.

Sometimes a pair of cells remain stuck together after they have divided in the sense of forming a partition between them. In a few cases anything up to a hundred cells may form a "colony." If they are all alike, this is not very interesting. But occasionally some are specialized for feeding and others for reproduction. We have the very beginning of the aggregation and specialization of cells which we find in the higher animals.

3

Jellyfish and Polyps

YOU and I have one set of hollows in us communicating with the outside by the mouth. We have others which open by other orifices, and several, including the blood vessels, which do not open to the outside at all. A bee or a snail has the same relatively complicated structure, and so do many simpler animals, such as earthworms.

But the three simplest main groups of many-celled animals, or phyla, only have one set of hollow organs. Perhaps the simplest animals, and certainly some of the most ancient, are sponges. A sponge is a mass of cells traversed by small channels, which lives by filtering water through itself and catching suspended particles. A bath sponge is the skeleton of one kind. Others produce chalky or flinty skeletons which are not used by man. There are several different kinds of cells in a sponge, and some of them are extremely like one group of protozoa, so it is at least possible that the sponges are descendants of members of this group which stuck together.

The next most primitive group, apart from some microscopic parasites which may be degenerate, is called the coelenterates. They have a mouth and a stomach which communicates with different parts of their bodies. The most familiar forms to us are sea anemones and jelly fish, but the corals are much more important, as they have been in the past, and still are, builders of masses of rock.

A typical coelenterate has a number of tentacles with which it catches its prey, a mouth in the middle of them, through which it also rejects the undigested remains of its food, a body wall with muscles, and sexual organs whose products may burst through the skin or be shed through the mouth. It may be a swimmer like jellyfish, or a fixed polyp, like sea anemones and corals. More generally, it goes through both phases. An animal roughly like a sea anemone buds off a whole series of jelly-fish, mouth upwards, which swim off, grow larger, and produce eggs which give rise to tiny animals which settle down again on a rock.   A jellyfish has a nervous system consisting of a network of fibres which coördinate its movements so that the whole bell contracts at once, and is often connected with "eyes" which are sensitive to light though they cannot perceive form, and an organ to enable it to keep the right end upwards.

Can we say that such a simple animal has any feelings, or anything approaching consciousness?   I don't know, but I doubt it for this reason.   Our own intestines have a nervous system rather like that of a jelly fish which enables them to contract rhythmically as it does.   Now if the nerves connecting our intestine to our brain are cut, we do not feel a stretching of it which would be very painful if the nerves were working normally. It is perhaps possible that there is a pain in the intestine, with no one to feel it.   But unless this is so, I think it rather unlikely that a jellyfish feels pain or pleasure.

The higher animals probably evolved from coelenterates, but the coelenterates also tried a method of evolution of their own, which did not get them very far.   In some species a group of animals like simplified sea anemones divide, but do not separate, and each one is specialized.   So a single "colony" can consist of one animal specialized as an air bladder for floating, several for swimming by jet propulsion, others for protection, a number for eating, and finally others for breeding.   They all share a common digestive tube, and between them constitute a very original sort of animal.   One of these colonies has even developed a sail, by means of which it drifts before the wind. This is one of the many lines of evolution which has been tried out by several groups of animals, but has never produced an animal as integrated as a fish or a beetle.

Closely related to the coelenterates, and often included among them by classifiers, are a group called the ctenophores, or comb-jellies, which are now usually put in a separate phylum.

They have the beginnings of a third layer of cells between the skin and the stomach lining, such as exist in all higher animals. They also have a pair of excretory pores apart from the mouth.

Most of them are swimmers, with up to six fin-like organs which give them their names. A few species are quite common in the British seas. But the most remarkable ones have taken to crawling on the sea bottom, and begin to look like the simplest worms. What is more, their development is rather like that of some worms, and they may possibly resemble the ancestors of worms, and hence those of higher animals. However we do not know whether they are a primitive group, because none of them have shells or skeletons which have been fossilized. The coelenterates are certainly a very ancient group. And it is quite likely that our ancestors passed through a stage of this kind before they developed heads, limbs, hearts, and so on.

Most coelenterates have stinging cells, especially in their tentacles, which enable them to paralyse small animals which come into contact with them, and to hold them while they put the tentacles into their mouths. They stay fixed or drift about aimlessly, eating food with which they happen to come into contact. The fact that they live shows that this is a fairly successful way of life. It enables them to get along with very feeble muscles, a very simple nervous system, and very incomplete information about their surroundings. They are very much at the mercy of their environment, they cannot search for food or migrate when conditions get bad. But on the other hand they have only to make skin, stomach lining, a jelly which consists mainly of sea water, and sexual products, out of their food. Their most successful members are the corals, which live in surf where the waves are constantly bringing them fresh food. Very few of them have managed to colonise fresh water, and none live on land. In fact they are pretty incomplete animals, and just for this reason intensely interesting for the student of evolution.

4

Worms

SIXTY years ago zoologists classified a great variety of simple animals as worms. To-day they divide them up into about fifteen phyla, each of which is thought to be about as different from the others as say snails are different from flies or fish. Nevertheless there was some sense in the old classification. Any animal more complicated than a coelenterate has various organs between its skin and its gut. A great many of them are fairly long, with a mouth at one end but no very definite head. From these, three different groups, which mostly have heads, have evolved separately, namely the molluscs such as snails, the arthropods, such as insects, crabs, and spiders, and the vertebrates, such as fish and men. The worms represent a large number of different phyla, or main lines of evolution, which have neither developed heads, hard shells, or various other specialized structures.

The simplest worms are the Flat Worms, which have organs between the skin and gut, but no hollow ones shut off from the outside like our blood vessels, nor what is called the coelom, the body cavity in which our internal organs, such as the heart and stomach, lie. Some of them are elegant little creatures, often black, which you can find in brooks under stones and leaves. But the most successful flat worms live in other animals as parasites. They include the flukes and the tapeworms.

The flukes, such as the sheep liver fluke, still have mouths and other organs. The tapeworms are extremely degenerate. They have no mouths, but generally live in the intestines of other animals, soaking in the digested food through their skins. Both these classes generally live in two hosts at different stages of their lives. For example the sheep liver fluke hatches out in water, bores its way into a snail, multiplies there, bores its way out, and attaches itself to a plant. If a sheep happens to eat this plant, it makes it way into the sheep's liver, and lays eggs which come out in the sheep's dung.

One of the tapeworms which infest men starts life in a pig, and only gets into the human gut if a man eats pork which has not been properly cooked. Others start life in fish. We can avoid them by seeing that our food is thoroughly cooked. Some worms have only one host, others as many as three. A tapeworm can live for years and produce many millions of eggs. On an average only one will give rise to another tapeworm, but that one has found such a favourable environment that unfortunately for us the system works. Another whole phylum of worms, the Acanthocephala, are all parasitic, but luckily they are rare in men.

A much more important phylum, the Round Worms, includes a great many free living forms which are found in soil, and are mostly too small to see clearly without a microscope. Those which eat the roots of plants are quite serious agricultural pests. But perhaps most species of round worms, and certainly the largest, are parasites. Most of us harbour one or two in our insides at one time or another in our lives, and are slightly ill in consequence.

The most formidable to man is the hook worm *Ankylostoma*. It was this animal rather than General Grant which won the American Civil War. The Negro slaves from West Africa harboured these worms. They spread if human excrement containing their eggs is left lying about, so that the larvae get out, and come into contact with the human skin. They burrow through it, and find their way into our intestines, where they suck blood and cause anaemia. The slaves had no lavatories and no boots, so they became heavily infected, as did their white masters. *Ankylostoma* needs warmth and moisture for its larvae. So the only serious outbreak in Britain occurred in the Cornish tin mines. It was at once controlled by providing the miners with proper sanitation.

A much more respectable phylum of worms is called the Nemertines. They are mostly sea animals, and about forty different kinds live round British shores, under stones and in the mud. They are lively muscular creatures, with a thread-like proboscis which is generally kept inside out in their bodies

K

but can be shot out after prey.   They are extraordinarily thin for
their length, which may be anything from about half an inch
to thirty yards.   In fact one of them, *Lineus marinus,* is the longest
of all animals.   A large whale only runs to about twenty-five
yards, and the worm can probably add another five or ten with
its proboscis.   But it is less than an inch in girth.   The nemertines
need a heart to supply oxygen to their vigorous muscles, and have
even got red blood, and, for worms, quite a good nervous
system.   I am glad to say that very few of them are parasitic,
and only sorry that none of them live on land in Europe, as some
do in the tropics.

I must pass over half a dozen or so small groups, each with
a few dozen species at most, but so unlike as to be regarded as
distinct phyla, to come to the most advanced worms, the Annelids.   These are divided up into a number of segments, each
segment having a structure rather like its neighbours.   One can
see that the skin of an ordinary earthworm is divided into rings.
This division is not merely skin deep.   Each segment has its own
section of the body cavity, its own muscles, blood vessels,
nerves, kidneys, and so on.   A few segments also contain sexual
organs.

The earthworms have bristles which point backwards and help
it to force its way through the soil, but nothing resembling legs.
However some of the sea annelids have a pair of appendages
on each segment which may be regarded as rudimentary legs,
and the appendages at the front end may be modified into
jaws or feelers.

Some of these worms also have eyes, so that naturalists have
dignified their front ends with the name of heads.   There is every
gradation from a fairly hopeful head to the condition found in
the earthworm.   This very fact makes it clear how a head can
evolve, by the concentration of jaws, eyes, feelers, and so on,
at the front end of the body.

The annelids include the species of worms best known in
Britain, namely the earthworms, the lug worm which is used for
bait, and the leeches.   They also include a number of very
fantastic sea animals, some of which are hairy like mice, others
live in tubes of sand grains stuck together, while yet others are

vigorous swimmers. An annelid worm with its many segments each with a pair of legs, its feelers, its jaws, its head, and so on, is not very unlike a centipede or a caterpillar. It is entirely probable that the insects and their relatives are descended from segmented worms. But we must wait for more fossils to prove or disprove this theory.

Another type of animal of which there are a number of phyla is the polyp, that is to say a fixed animal with a number of tentacles round its mouth, and occasionally a beak. The polyp's way of life, like that of the worm, has been adopted by animals of very different structures, some for example with hearts and gills, though mostly without. The most successful phylum of this kind, called the Polyzoa, have done their share of rock-building, though less than the corals.

5

Living in One's Skeleton

ONE reason why one regards a crab or a dog as a higher sort of animal than a snail or a starfish is that the crab and dog have jointed limbs with hard skeletons, capable of very precise movements. But there is a very big difference. The dog's skeleton is inside, the crab's outside. The group or phylum of animals which mostly have jointed limbs and external skeletons is called the Arthropods. Clearly their skeleton gives them a great deal more protection than our skin. But it has a big disadvantage. They can only grow by moulting, and are very vulnerable during the moult. The Arthropods fall into four main divisions; the Trilobites, sea animals of the general appearance of woodlice, and all extinct; the Arachnids, including spiders; the Crustacea, including lobsters, and the Progoneata, including centipedes, millipedes, and insects. Some writers include a fifth group for a strange animal called Peripatus, resembling a caterpillar which never becomes a butterfly.

An arthropod is built up of segments, some of which may be fixed, and each of which may carry a pair of appendages. The arachnids differ from the rest in having clawed appendages, instead of feelers, or antennae, on the front segment. They include such well-known creatures as spiders, scorpions and mites. They also include extinct sea animals up to six feet long, and the still living king crab, which measures a foot or so. But their queerest members are the pycnogonids, sluggish sea animals consisting almost entirely of legs attached to a mere vestige of a body. Even their digestion has to be done in the legs, and a tube extends down each leg from the stomach.

The crustaceans include a great many simple creatures like the water fleas; and some of the smaller sea forms are extremely successful, and very important as the main food of some fish which we in turn eat. Other crustaceans have developed into the complicated shrimps, lobsters, and crabs, where some limbs have been specialized as feelers, others as jaws, as pincers, as legs, and as swimming organs. Only a little less complicated are the sandhoppers, and the woodlice, which have established a foothold on land. But some crustacea start life respectably enough, as little swimmers like water fleas, and then degenerate, becoming parasites inside other crustacea or fish, and end up as shapeless lumps. An odder fate has befallen the barnacles. They stick themselves to a rock by their front ends, and live by kicking food into their mouths with their back legs. Meanwhile they produce a shell, not of horn, but of lime, which protects them against enemies and low tides.

None of the animals so far described have definite heads. A spider or a lobster has a front part with the mouth, eyes, and legs, and a back part or abdomen. But the insects, and their relations, such as centipedes and millipedes, have a head movable apart from the segments which carry the walking legs. The insects have six walking legs, one pair of antennae, three limb pairs for mouth appendages; and may or may not have wings. The other groups included with them, such as centipedes, have more legs.

The insects are, at least as regards numbers of species, incomparably the most successful of all animal groups. This merely means that each species is generally well adapted to a particular sort of life, and to no other. So there is room for an immense number of different insect species in the same area, not competing directly with one another. And among the insects the most successful order is the beetles. At the present time nearly a million species of insects are known, at least a third of which are beetles. And these numbers are growing steadily. A naturalist recently described over 400 new species of weevils (small beetles) from a single Pacific island. As compared with this there are only about 8,000 species of mammals altogether, and only sixty in Britain.

The insects are divided into a primitive group which have no wings and whose ancestors never had them, and a group whose ancestors at least have had wings, though some, like the fleas, have lost them. The primitive ones, such as springtails and "silver fish" are mostly small, though some play an important part in the soil.

The winged insects are not so clearly divided into those which grow up by a series of moults at which every stage is an obvious insect, and those which have larvae quite unlike the adult. Examples of those whose young are fairly like the parents are grasshoppers, cockroaches, and earwigs. Even the young of dragonflies and mayflies, which live under water, are recognizable as insects.

But no one would guess that a wasp grub, a fly maggot, a swimming mosquito larva, a "wire worm", or a caterpillar would grow up into a winged form, unless they had some knowledge of biology. The larval forms have learned to live in all sorts of environments. Thus some are burrowers. One lives underground for seventeen years, followed by a week or two as a flier. Others live in decaying wood or meat. Still others are parasites inside other insects or occasionally in mammals. But this sort of parasitism does not lead to degeneracy. The adult female of a species which grows up as a parasite in a caterpillar has to have very well developed senses to find just the right kind. She does not rely on chance, like the tapeworm, to find a host.

Still more remarkable, a parasite can evolve away from parasitism. The most primitive members of the order Hymenoptera lay their eggs in plants, where the larvae live in galls, or else in caterpillars and grubs. The next higher group catch grubs which they paralyse by stinging, and on which they lay eggs. At a later stage, some wasps feed their young with chewed up animal food. And finally the bees feed them with pollen, and with their own secretions. This development of parental care and social life from parasitism is one of the most amazing in the whole history of evolution. It may of course be compared with the evolution of human society through slavery and class oppression to communism.

Roughly speaking the insects on land and the crustaceans in the sea fill the same kind of place, the place available for moderate-sized animals with highly developed senses, complicated instincts, and some intelligence. Luckily for us, they cannot grow very large. Imagine a cow which had moulted its whole skeleton, and was waiting to grow a new one, and you will see why. But it is perhaps mainly because of this limitation that the arthropods have not completely conquered the world.

6

Shelled Animals

A NUMBER of animal phyla have developed shells mostly made out of lime, and live inside them. The most successful of these phyla is the molluscs. They are fairly complicated animals, with hearts, gills or lungs, kidneys, livers, glands of internal secretion, and so on.

There are three successful classes. The gastropods, of which the snail is the best known, have what may be called a head, and typically a "foot" on which they crawl. Their shell is generally coiled, and they are the highest animals except the flat fish which are quite asymmetrical. However some of them such as the limpets, sea slugs, and the beautiful swimming snails called pteropods, have straightened out again externally, though they still keep a little asymmetry inside. The snail plan of life is a pretty successful one, for numerous snails live in fresh and salt water, and on land, and a few slugs burrow quite deep in the earth, using their small shell to push the earth aside.

The bi-valved molluscs or lamellibranchs, like the mussel and oyster, have specialised in defence, and do not move very fast, though cockles can jump and scallops can swim. They mostly live by filtering small particles out of water. They have never come out on land, and have not even colonised fresh water on a great scale.

The third great class of molluscs, the cephalopods, such as the squid and octopus, have eight or ten tentacles round their mouths, and a very definite head, with excellent eyes and other sense organs. In the past they included the very successful ammonites, which lived in coiled shells, but to-day almost all of them have internal shells which merely stiffen them, or none at all. They all live in the sea, and are probably the most intelligent of sea beasts. The octopus has a high degree of control over its tentacles, and uses them for building a nest of stones.   Among the cephalopods are the heaviest of all invertebrates, huge squids weighing many tons and measuring up to fifty feet in length, including the tentacles.   But the cephalopods seem to be a dead end.   For hundreds of millions of years they were among the most numerous sea animals.   To-day they are comparatively rare except in the Antarctic seas.

There are two minor classes of molluscs.   One has a shell divided into eight plates, and looks rather like a woodlouse. The other lives in long tubes like miniature tusks.   The molluscs are handicapped for movement by their clumsy shells.   A number of them, like the slug and the cuttlefish, have given them up or only kept them as stiffening.   But an animal without a skeleton is handicapped too.   So on the whole they have been less successful than the arthropods or vertebrates.

Another phylum called the brachiopods make bivalve shells which are pretty like those of the bivalved molluscs, though their soft parts are very different.   They have made less progress than any animal group known to us.   One type, *Lingula,* has evolved so little that the living shellfish are assigned to the same genus as those which lived four hundred million years ago. That is to say they do not seem to differ much more than horses and donkeys, or cats and leopards.   No wonder the brachiopods are very much rarer to-day than they were in the past.

Another great phylum of animals, some of which have shells, is called the echinoderms.   Some are fixed to the sea bottom, others move about slowly.   The most typical members have a five-fold symmetry like a primrose or wild rose flower.   The best known echinoderms are starfish and sea urchins.   They are unlike all other animals in quite a number of ways.   For example many of them have a system of tubes full, not of blood, but of sea water which they use for expanding hollow "tube feet" which they can force out through holes in their shells.

The sea urchins have also solved a problem which no other animal has solved, namely how to live in a box and yet to grow. The snail has one end of its box open, which it can sometimes close with a shelly door.   The mussel has two valves, but these can be forced open.   The tortoise is open at both ends.   The sea urchin's shell consists of a number of plates which grow at their edges, but cannot be forced apart. It has a small, well protected mouth, and numerous pores through which tube feet, and muscles to move its spines, protrude.

The earliest echinoderms were probably fixed to the sea bottom. Most of the fixed types are extinct, though some still live, looking rather like plants. In fact they are called sea-lilies. In several of these the "flower" breaks off the stalk when it is fully grown, and swims away as a kind of starfish. However the ordinary starfish and urchins are never fixed, whatever their ancestors may have been. Just as in the molluscs, there is a conflict in the echinoderms between the needs for movement and protection. Some, like the starfish, have a tough skin with spines, but no box. Others, like some of the sea cucumbers, have quite thin skins.

We do not yet know much about the interrelations of the main phyla of animals, largely because all except the vertebrates had already developed their special characters five hundred million years ago, when the Cambrian rocks, the first which contain many fossils, were laid down. But their development tells us something. Molluscs often start life as a swimming larva startingly like the larvae of some annelid worms. They are probably more nearly related to them than to the echinoderms or the round worms, for example.

And strangely enough, in their early development the echinoderms pass through a stage with only one plane of symmetry, not five, which is not very unlike an early vertebrate embryo. They also resemble the vertebrates in several chemical matters. If these two phyla are related, the connecting link may be a group of extinct animals called the Machaeridia which died out three hundred million years ago, and were like echinoderms, except that they had one plane of symmetry, instead of five. Nobody, of course, suggests that our ancestors were sea urchins. But it does seem likely that one would not have to go back quite so far into the past to find the common ancestor of men and sea urchins as the common ancestor of men and bees or of men and snails.

7

Fish

WE men belong, together with four-footed animals, birds, fish and so on, to the phylum called Vertebrates, characterised, among other things, by a many-jointed back-bone.

There are a number of groups of animals, some of which are fairly surely, others more doubtfully, related to the vertebrates. A worm called *Balanoglossus* has a rod resembling the horny rod which precedes the backbone in our development, and slits in its gullet like the gill slits of a fish. It also agrees with us and with some of the echinoderms, and differs from most other invertebrates, as regards the biochemistry of its muscles. Some zoologists take these resemblances more seriously than others. If it is a relative, so probably is a living creature called *Rhabdopleura* which lives a life rather like a tiny sea anemone in the deep sea. And so were a large extinct group of polyps called the graptolites.

Few zoologists doubt that we are related to the tunicates or sea-squirts. When adult they are mostly fixed to rocks, sucking water in through one hole and squirting it out by another after filtering food out of it. But many of them start life as "tadpoles" resembling a vertebrate in a great many respects, and some of these tadpoles do not settle on rocks, but swim throughout their lives. Roughly speaking we may say that the tunicates are related to fish and men as barnacles are related to shrimps and bees. They have become fixed and lost many organs. We have kept moving and gained new ones.

A much nearer relative is the lancelet, *Amphioxus*, which may be described as a fish without a head. As it is also boneless the relationship is hard to prove from the fossil record. But by great good fortune a few quite ancient fossils of a boneless animal called *Jamoytius* have been preserved in shale, and are like lancelets but with better developed eyes. So the few who doubted the relationship of *Amphioxus* to the vertebrates are now fewer.

The most primitive living vertebrates are the lampreys and their relatives. Superficially they look like eels. But they have no paired fins and no jaws. Instead they have a round mouth and horny teeth, with a number of gill slits opening out of their gullets. The most ancient fish whose skeletons we possess resembled the lampreys in having no jaws and no paired fins, but were heavily armoured. They were probably not the ancestors either of modern fish or of men. The next group of fish to appear on the fossil record had primitive jaws and were beginning to develop paired fins. Different lines of them tried one, two and three pairs of fins, and only those with two pairs have left descendants.

The fish which are alive to-day belong to many groups, but only two of them are important. The sharks and rays have skeletons of cartilage, not of true bone, and are primitive in many ways.

They have probably survived because they look after their children better than most other kinds of fish. Some, like the dogfish, lay large eggs with tough shells and plenty of yolk. Others like some rays, bear their young alive. So they produce only a few children, but these are of a fair size when hatched or born, and each stands a good chance of growing up.

The more modern type of bony fish, though they are more advanced than the sharks in many ways, usually lays great numbers of small eggs. So there is a huge infantile mortality, and very few young fish live to maturity. They have however been extremely successful, largely because they can swim quicker than any invertebrates, and though most of them are much alike, some have specialized in extraordinary ways.

Several groups have taken to lying on one side, and are almost as asymmetrical as snails. It is well worth looking carefully at the next flat fish you or your wife buy. And if you don't believe that it was evolved from a symmetrical ancestor, it is up to you to explain why it hatches out symmetrical, and then takes to lying on one side, while its eye moves round to the other. Others, like the eels, have lost, or nearly lost, their paired fins.

Some have taken to living in bony boxes, often spiny, like echinoderms, and can only swim very slowly.

Those which live in the great depths of the ocean are often highly specialized. Some have luminous organs with lenses, which enable them to see in what would otherwise be utter darkness. Others which live in the middle depths a mile or so below the surface and above the bottom have immense mouths, and probably only get one or two meals a year when some dead animal falls from near the surface. One, I regret to say, lives inside a sea cucumber. This is the only example of parasitism in an adult vertebrate. But another fish, the bitterling, lays its eggs in the gills of a fresh water mussel, and is therefore a parasite when young.

There are still a few living fish which are fairly like those which went ashore in the Devonian, or Old Red Sandstone Age, and became our ancestors. They have paired fins with a bony axis which is much more like the limb of a four-footed animal than are the spiny-fanned fins of modern fish.

A great many of the bony fish have air-bladders. Some can rise to the surface to fill them and thus supplement their gills if the water gets foul. Others have closed air-bladders whose only use is to buoy them up in the water. Among the living fish which are most like our ancestors are the lung-fish of Australia and Africa, which burrow into mud when the swamps where they live dry up, and can breathe air for months on end.

Several living fish have paired fins modified for walking, either on the sea bottom, like the gurnets, or on land, like the mud skippers, and the evolution of some kind of limbs was not the biggest problem which faced our ancestors when they came out. One of their problems was reproduction. Most female fish lay eggs in the water, the male pours sperm over them, and the eggs are left to themselves. The first land vertebrates doubtless had to go back to the water to breed, and there are still plenty of living animals such as frogs and newts, which have not yet got over this necessity.

8

Beasts and Birds

LIFE began in the water. The first land animals of which we have fossils are flightless insects, presumably descended from something like the modern crustaceans. Later, perhaps fifty million years later, fish came on land and developed legs.

Perhaps it was to get away from them that insects took to flying at the time when the coal was formed, or maybe earlier. About the same time the first four-footed animals produced eggs with shells tough enough to prevent the young from drying up, yet sufficiently porous to allow oxygen to soak in. Once this was done, they did not need water to breed in, and could live anywhere on land, except perhaps in very cold regions. We speak of these early four-footed animals as reptiles. But I have little doubt that in another fifty years or so we shall use a different classification.

Like any group of animals which conquers a new habitat they specialized in all sorts of ways. One group, from which we are descended, quite early developed the three different kinds of teeth, cutters, dog-teeth and grinders, which are found in most mammals. And they probably had hair. They seemed to be the most progressive animals. Yet they mostly disappeared, and were ousted by a group called the Archosaurs (ruling reptiles) which dominated the land for a hundred million years. Their best known members were the dinosaurs and the flying reptiles, or pterodactyls. Their living descendants are the birds and the crocodiles. A great number of them walked on their hind legs and used their fore paws for grasping. The one thing they did not do was to develop their brains.

The existing lizards and snakes are only distantly related to them, being descended from a group which was not very important during the great age of reptiles. The tortoises, which have gone in for armour and slowness, like sea-urchins,

snails, and crabs, are even more distantly related.   At least seven
and probably many more groups of reptiles went back to the sea,
and of these the ichthyosaurs were as highly specialized as the
modern whales, even bearing their young alive.   Of these groups
only the turtles and sea-snakes survive.   Two or perhaps three
groups learned to fly, of which only the birds survive.   Others
developed gigantic size, and some of the plant-eaters had more
efficient grinding teeth than any living animal.

In fact during the time between the formation of the coal and
the chalk, nature tried a vast variety of experiments with four-
footed animals.   About the end of the chalk, almost all of these
came to an end.   We are absolutely in the dark as to why they
did so.   There are any number of theories, one as good—or bad—
as another.

Fortunately the ancestral mammals, descended from a group
of reptiles who had almost disappeared before the archosaurs,
were there to take their place.   There was such a vacancy for
large land animals that quite a number of flightless birds, like
giant toothed penguins, stepped into it.   But they were soon
ousted by the mammals, which have now had seventy million
years to adapt themselves to various ways of life.

They have done most of the things which the reptiles did,
on a higher level.   Like the pterodactyls, the bats have taken
to the air; like the ichthyosaurs, the whales have assumed the
form of fish.   The moles are more efficient burrowers than any
reptiles of which we know.   On the other hand the kangaroos
are the best mammalian attempt at a pattern of animal resting on
its hind legs and tail which was rather common among the dinosaurs.   And the armadillos are not so effectively armoured as
the tortoises.   We mammals have not produced a form like the
snakes; and though in each age there have been a few monstrous
forms like the elephant and the rhinoceros, the reptiles certainly
evolved more and heavier monsters, apart from the whales.

When we think of mammals we probably think mostly of
hoofed animals like the cow, deer and horse, carnivores like the
lion and wolf, and animals with hands like the monkeys and
ourselves.   But it is striking that about half the living mammals
belong to the order of Rodents, which includes rats, beavers,

and porcupines. On the whole these are the most successful and wide-spread of mammals.

I think the most original mammals are whales, elephants and men. The largest whales are all filter-feeders. The largest reptiles ate plants or fairly large animals. But the great whales have no teeth. They live on shoals of shrimp-like crustaceans which they strain out of the sea with their sieves of "whalebone" which have replaced their teeth.

The elephants have developed a trunk which allows them to do at least some of the things which we do with our hands, while using all four legs for walking. If they had taken to doing something more constructive than pulling down branches with it, if for example the mammoths which lived in cold countries had started making houses to keep out the snow, they might have developed their brains as we have.

Man is rather a primitive mammal as regards structure. He has kept all his fingers and toes, and most of his teeth. He has not got a highly specialized stomach like a sheep, or a new organ like an elephant's trunk or even a cow's horns. Indeed the part of his body which is most different from the corresponding part in his nearest relatives, such as the gorilla, is probably his heel, which enables him to walk for long distances on one pair of legs.

His unique features are his very large brain and the use which he makes of it. He is not the only animal that uses tools. There is for example a bird which uses a cactus thorn to pick insects out of holes, not to mention spiders with their webs. He is not the only animal which builds. Most birds do so. Many animals store food, and some ants domesticate other insects. But he is the only animal which deliberately shapes tools, and the only one which uses fire. Nevertheless he was a fairly rare animal till quite recently, and it is only in the last twenty thousand years that he has lived in societies bigger than large families.

As men learned to co-operate, a whole series of new problems arose, and men without ceasing to be animals passed out of the sphere of the science of zoology, just as living things, without ceasing to be material, passed out of the sphere of the science of chemistry.

VII

ZOOLOGICAL ESSAYS

I

Why Steal Beetles?

AN entomologist has recently been sent to prison for stealing beetles from the British Museum. I do not know whether he stole them because he liked them or because he intended to sell them. The value of those stolen was said to be several hundred pounds.

One may well ask why beetles should be valuable. Some sorts are regrettably common. But there is always one particular beetle, even of the commonest kind, which is specially precious. This is the type specimen, on which the person who first described the species based his description, and it generally exists in some museum or other. The zoologist who named the species may not have given a very accurate account of it. But provided the type is available, and undamaged, others can give a better one. Why does this matter? Why is it any more important than the description of a kind of postage stamp?

The reason is this. If we know to what species an animal belongs, we may know a great deal about it. It is important to know whether a beetle is a scavenger or eats wood, crop plants, or something else of value. Even if it can eat potato plants it is not a menace to British agriculture if it belongs to a tropical species whose members are killed by a mild frost.

Of course within a species there may be several geographical races differing in their resistance to cold, and what is more striking, in thirty years or so a race may arise which can eat a new food plant.

Nevertheless most species are pretty well adapted to one sort of environment, and find it rather hard to live in another one. This is particularly true of beetles. We do not know how many species of beetles there are. But there are certainly more than four hundred thousand. That is to say something like a third of all the known species of animals are beetles. However, beetles are not enormously more successful than other insects. The total number of individuals of Collembola, a group of insects which one can hardly see without a lens, is probably much greater than that of beetles. The reason why there are so many beetles is probably that they tend to be specialists, for example eating only a particular fungus which they find under the bark of a particular kind of tree.

Now the distinction between species is made on the basis of characters such as the number of bristles on their legs, which have no obvious importance. They are important because they enable us to assign an animal to the correct species, just as the details of my face are important because they enable people to recognise me.

Taxonomy, that is to say the assignment of animals to species and of species to genera, families, and so on, is only a part of zoology, but it is an important part. Darwin called his greatest book *The Origin of Species*, and he knew what he was talking about, because he had spent eight years in classifying living and fossil barnacles.

Since Darwin's time we have learned a very great deal about another aspect of evolution, namely the slow changes by which, for example, the descendants of three-toed short-toothed animals have become horses with only one toe per leg, and long teeth suited for chewing grass. We can learn about this from fossils. But fossils do not tell us anything about the kind of difference which exists between a horse and a donkey, and causes the hybrids between them, the mule and jennet, to be sterile. This is a very important difference, and we can only learn about it by studying living species, and geographical races, which seem sometimes to be species in the making, because they may give sterile or weakly hybrids.

There has been a natural, but unfortunate, tendency to confine taxonomy to museums, so that university students learned very little about it. London University has just appointed a reader in taxonomy to combat this tendency. I do not envy him his job. He too will need a large collection of beetles or whatever small animals he may use, for our university has neither the space nor the money for a collection of animals as large as deer or even birds. He will have to teach students how to distinguish different species, and how to decide whether an animal belongs to a new species previously undescribed. And no doubt he will explain that the needs of a species, as shown by the food which it eats and the climates in which it can live, are even more important than the characters used for describing it.

The practical importance of such work can be judged by a simple fact. The Government has great schemes for growing groundnuts and sunflowers in tropical Africa. It is quite useless to grow a plant which needs insects such as bees to fertilize its flowers in an area where there are not enough of the necessary insects. It is equally useless to do so where there are insects which will devour the crop. I do not know whether there has been an adequate survey of the insects in the African areas concerned. If not, the whole scheme may fail for this simple reason. It may seem ridiculous that the measurement of bees' tongues under a microscope could decide whether sunflowers can or cannot be grown profitably in a particular area. I hope it did not seem ridiculous to the people responsible for these schemes. If it did, the results may be pretty serious. A vast amount of public money will have been wasted, and we shall have less margarine in 1950. But this simple example shows that a collection of insects has a use value as well as a rarity value like postage stamps.

Of course the majority of insects have no economic importance, and are not likely to have one. But you do not know this beforehand. It would be short-sighted and impracticable to try to ignore animals not known to be of economic importance. Darwin's barnacles are a good example. If we knew how to stop barnacles growing on ships' bottoms we could save a vast amount of coal and oil. Among the things which Darwin discovered in barnacles was the apparatus by which they cement themselves on to rocks or ships. Certain kinds of paint make it harder for barnacles to do so. But none of them is fully efficient. It may need a man with Darwin's powers of observation and insight to solve the problem completely. The novelist Lytton Bulwer satirized Darwin for his interest in barnacles. To-day Dr. C. D. Darlington thinks that too much of the national income is being spent on taxonomy, especially at Kew.

Taxonomists can of course become narrow and unduly academic. So can scientists of other kinds. But taxonomy is of great practical and theoretical importance, and the beetles in the British Museum are a really valuable national property.

2

Cage Birds

I HAVE just been to the annual Cage Bird Show in London. Some people at once object to such a show because it is cruel to put birds in cages. This is certainly true for migratory birds such as swallows, and probably for birds which go in for long and high flights, such as falcons or skylarks. And certainly exhibition cages are rather small. But the two most popular species, the canary and budgerigar, are very tame, and often do not fly away when they are given the chance. Neither would live very long in England if it did.

I went to the show mainly as a teacher, because a lot is known about the inheritance of characters in the budgerigar, and a good deal in other species, and because the characters bred for are quite superficial compared with those, say of milk or beef cattle, so that one can easily show them to a student. I write about it here, because most breeders of cage birds are workers, and I hope that as hours are shortened and more houses are built, more and more workers will keep some sort of pet animals, if only for the sake of their children.

The budgerigar is one of the most interesting domestic animals, because we know its history. It lives wild in north Australia, and the first live ones were brought to Europe in 1840. They were light green, and since then a great many other colours have turned up. There are many shades of blue, yellows, greys, and whites, as well as light green, dark green, and olive. Also there are various combinations of these colours, such as greens with yellow head and wings, and variations in the pattern of stripes on the neck. So far there have been no important changes in structure, such as are found in pigeons, for example the pouter with its expanded chest, the fairy swallow with its feathered legs, or the fantail, whose name needs no explanation. A crested form is said to exist on the continent, but was not on show in London.

In other birds the breeds are generally kept sharply separate. Indeed the pigeon fanciers' association have very remarkable names such as the Oriental Frill Club, and the Bald and Beard Club. But the budgerigar fanciers know enough genetics not to be afraid of crossing their breeds. They know for example that if their strain of mauves is getting rather weak they can outcross to the wild coloured light green, giving dark green young, and that if these are crossed to mauve about half the chickens will be mauve. Moreover, these new mauves are likely to be invigorated as the result of the outcross. The fanciers call this sort of mating "dipping into the green".

The canary fanciers, on the other hand, have a number of different races, such as the Norwich, Yorkshire, Border, and so on, which they try to keep "pure", though each race has several different colours. However there is one exception to this rather Nazi practice. If you mate a crested canary with a plain-headed one, you get about equal numbers of crested and plain chicks. If you mate two crested together you probably get more crested than plain among the chicks which live, but a large number of your chicks die before hatching. There is no way of getting crested canaries which breed true, and in practice they are out-crossed to specially bred consorts. Several characters in mice and poultry behave in the same way.

I wonder how many of the inborn human characters which we regard as desirable show this type of inheritance. Nobody knows, but it is far from sure that even if a dictator had absolute power to decide who should have children by whom, he would get the best results from his point of view by mating like and like.

Animal breeding may have two distinct objects; to produce better animals, or at least animals which will fetch a high price or win prizes; and to increase knowledge. The fancier cannot be of great help to the farmer who breeds animals as producers of food or wool, for a very simple reason. The farmer wants forty cows all of which are good milkers. The fancier may breed forty canaries, and if one of them wins a national challenge trophy he does not much mind if the other thirty-nine are second-rate birds. His aims are quite different to the practical farmer's.

But the fancier can advance science, and in this way he can help agriculture indirectly. For example the principles of sex-linkage were discovered in canaries before poultry, but they have been of great value to poultry breeders.

A group of cage bird breeders might quite well decide that it was as interesting to breed for knowledge as for silver cups. If they did so there is a very great deal that they could find out. There is plenty to be discovered about inheritance in budgerigars, but far more in canaries. And we know almost nothing about the various kinds of hybrids between canaries and related birds, such as linnets, goldfinches, and green-finches. We know for example that such crosses produce about ten times as many cocks as hens, and that some of the hens look much more like canaries than any of the cocks. We know that most if not all of these hybrids are sterile. But we don't know why in any detail.

We know that some budgerigars can talk. One which was shown produces the remarkable sentence "Johnny Stockdale, Rooks Hill, Welwyn Garden City", which would be very useful if he were lost. But we do not know if the ability to talk is inherited, let alone how it is inherited.

Plenty of amateurs have made real contributions to science, particularly to archaeology, geology, and meteorology. Bird fanciers could do so. One essential is to record all your animals, not only those which come up to some standard. Another is to breed fairly large numbers, because many of the laws of inheritance are numerical. This means that co-operation is essential. But as a knowledge of science is spread among the people of this country, more and more should be interested in advancing it and willing to work together to do so. Few people could make more solid contributions to science in their spare time than the breeders of cage birds.

3

Some Queer Beasts

I VISIT the London Zoo fairly constantly, but I suppose my taste in animals is rather different from most readers'. Still they may be interested in a scientist's tastes, even if they don't share them.

If I had to pick the most striking animal on show, I think my vote would go to a small fish called the mud-skipper. It lives in tropical mangrove swamps, and spends most of its time out of the water. At any rate those in the Zoo aquarium do so. The mud-skipper has fins, but at least one pair of them have a joint like an ankle, which enables it to use them for a clumsy kind of hopping on land. Its eyes bulge out of its head, and even move up and down like a frog's. Have you ever watched a frog eating? It has no complete roof to its mouth, so its eyes move up and down when it eats. In fact it uses its eyes to help it to push food down its throat.

I like the mud-skipper because he is trying to do what our ancestors did when they came out of the water in Devonian times. He is obviously a fish, and some of his near relatives are quite typical fish. But he has turned one pair of fins into passable limbs. He gives one an idea of what the first ancestors of the land vertebrates were like. If any critics of evolution think that fish could not have come out of water and become amphibians, you can show them the mud-skipper. Fortunately for him, he does not know that he is about three hundred million years late in his attempts to colonize the land from the water.

A fish which is interesting from a very different point of view is the Cichlid from the lake of Galilee. Not only is it almost certainly a fish of one of the kinds which the twelve apostles caught, but it was probably involved in a miracle. These fish have remarkable breeding habits. A pair of them scoop a hole in which the female lays her eggs. The male picks them up, and carries them round in his mouth for a week or more until some time after they have hatched.   As these fish also generally use their mouths to remove stones from the hole where the eggs are laid, they would be quite likely to pick up a small coin if one were lying on the bottom.   Now on one occasion the apostles are reported to have had no money to pay a tax, and to have got it by hooking a fish with a "penny" in its mouth.   It seems likely that this fish was caught while making a nest.

My favourite house at the Zoo is what is called the Temporary Rodent House.   It contains a number of rodents, which surpass all the other seventeen living orders of mammals both in numbers of individuals and of species.   I should explain that the class of mammals, that is to say warm-blooded hairy animals which suckle their young, is divided into thirty-two orders, such as elephants, whales and porpoises, bats, and carnivores, which include cats, dogs, bears, weasels and so on.   Fourteen of these orders are extinct.

But this house contains representatives of three orders which are not generally known.   The hyrax looks rather like a guinea pig until you look at its feet carefully, or better, dissect it.   It then turns out to be nearer to hoofed animals such as pigs. Actually it is fairly like the ancestral forms of many different mammalian orders about the time when the last of the chalk was being formed.   But it has evolved much less than most of the others.

Then there are the sloth and the tree ant-eater, representing the order called Xenarthra, or edentates, which originated in South America and of which a few members have got to Central and North America.   This order also includes armadillos, and used to include giant sloths about as large as elephants, and armadillos six feet long.   The ant-eater has no teeth at all, the sloth has very few and simple ones, and a very much worse temperature control than most mammals.

Finally there is the South African aardvark, the sole survivor of an order called the Tubulidentata.   It has a nose like a pig's, ears like a rabbit's, and is possibly the champion digger of the world.   It not only lives in a burrow, but gets its food by ex-cavating white ants' nests.   Whereas the mole chooses soft earth in which to hunt worms, the aardvark works in hard and dry soil, and does it very well. If its den had not got a stout cement floor, it would be out of sight in a few minutes.

The same house contains several galagos, which are small lemurs not unlike one of the forms ancestral to men and monkeys. Unfortunately they do not like light, and to see them at their best you must visit the Zoo at night, which will not be possible till we have more coal. But in the night time they are astonishingly active and graceful.

Naturally I don't expect other readers to share my tastes. I like to see a set of animals which illustrate the various possibilities of evolution, some of which have only rarely been taken. Of course some of the invertebrates have done much odder things, for example the hermit crabs which live in coiled shells, and have their bellies bent sideways to fit them, or the barnacles, which start life swimming about like little shrimps, and then glue their heads onto rocks and live by kicking food into their mouths with their back legs.

Every mode of life has its corresponding structure, and palaeontologists have a big task in trying to puzzle out how extinct animals lived from a study of their bones and teeth. On the whole they are pretty successful, but in one or two cases it is very hard to see how the animal worked. Perhaps it had some soft part of which we know nothing, like the chameleon's tongue, which he can shoot out for a foot or so to catch flies.

Some vocations are equally queer. If you haven't seen an aardvark you may find it hard to believe that there is such a beast. And if we didn't know there were such people as stock-brokers and tick-tack men we might not credit their existence either. Perhaps both may become extinct within a comparatively short time, I even venture to hope within my own lifetime. After all we manage without druids, rain-makers, augurs, exorcists, and quite a number of other professions which have been considered important enough in the past. It might be worth while keeping a few members of these professions if they were as odd looking as ant-eaters or mud-skippers. But their lives have not modified their structure. So let them go, provided we can keep the aardvark.

4

Counting Wild Animals

ONE of the first things we want to know if we are to make natural history scientific is how many animals of a particular kind there are in a certain area. We also need such knowledge if we are to use our land scientifically.

In a few cases the counting is fairly straightforward. If one wanted to know how many mussels there were between tide-marks on a particular beach it would be hopeless to count them all. But it would be quite easy to take a wire square covering just one square yard and throw it down on a hundred different spots, counting the mussels in each square. This "Gallup poll", with a measurement of the total area, would give the total number within ten per cent. or so.

It is obviously much harder to count animals which move about. Here again the sampling method may be quite useful with slowly moving animals like earthworms or beetles. One can dig up a square yard of meadow, and count all the earthworms in the turf and the soil beneath it, or in a fair sample of them. The results of such counts are very surprising. Not one person in ten has even seen the commonest British insects, the Collembola, which are wingless subterranean creatures about a millimetre long when fully grown, and whose numbers run up to four hundred million per acre.

It is not so easy to count more mobile animals, though curiously enough the exception proves the rule. Birds are the most mobile animals, but are easy to count in spring, because we can count their nests. For example, there are about 30,500 breeding pairs of rooks in an area of 910 square miles of the Upper Thames Valley, 4,000 pairs of herons in the whole of England, and a hundred thousand million birds in the whole world. This number may well be out by a factor or two or three, but not much more.

It is much more difficult to count mobile animals such as

flies or fish which have no definite nesting places, or whose nesting places are not easily discovered. Suppose we want to count all the field mice in a meadow, we might put down traps till we had caught them all, but before we had done so others would have moved into the vacant territory from outside. The only satisfactory method is to mark the animals. Field mice are best marked with a numbered nickel band round a hind leg, butterflies with a spot of coloured paint on one wing, fish with a leaden button at the base of one fin, and so on.

To count field mice, Hacker and Pearson put down six traps provided with food and bedding in every hundred yard square of an area of woodland. Each mouse was labelled at its first catch, and the fact that the same mouse was often caught several times a month made it clear that almost all the adult mice were caught at least once in a period of six months or so.

To count tsetse flies in Africa Jackson invented a rather different method which is however best illustrated by the work of Dowdeswell, Ford and Fisher, who counted all the Common Blue butterflies on the island of Tean in the Scillies. On each fine day they caught anything up to sixty butterflies, released them after marking, and saw how many of these they caught next day. Thus on one day they caught 52 butterflies and marked them all. Next day they caught 50, of which 14 had been marked the day before. That is to say 28 per cent. of the catch were marked. So if there had been no deaths or hatchings in the twenty-four hours, the total population was such that 52 was 28 per cent. of it. In fact there were about 186 butterflies on the island.

They could thus count the population on different days, and also by seeing how the proportion of marked insects fell off, discover the average length of life of a butterfly after it has come out of its chrysalis. No marked butterflies flew to the nearest island, about 300 yards away, so their problem was much simpler than Jackson's with his tsetse flies. Here it is necessary to find out how far a fly can move during its lifetime, and this distance may be several miles. There may be about ten thousand of these flies in a square mile of African bush. And as they can infect men with sleeping sickness and cattle with the fatal disease called nagana, it is very important to wipe them out.

A great many methods are being tried. They can be kept down to some extent by burning grass and bushes, and by killing off the deer, zebras and other animals on whose blood they live. Even trapping reduces their numbers slightly.

However, the most hopeful methods seem to involve the new insecticide D.D.T. In the South African Union there is an isolated area infected with tsetse near the coast of Zululand. This is being treated with D.D.T. smoke from aeroplanes over most of its extent, and from smoke candles in valleys where planes cannot penetrate. In other areas the cattle are being sprayed with D.D.T. so that the flies which settle on them die. However, unless not merely ninety nine per cent. of the flies, but all of them, are killed, these methods will be of little use. If half measures are adopted for some months, a race of tsetse resistant to D.D.T. is quite likely to evolve by natural selection, as a race of scale insects resistant to hydrogen cyanide has arisen in the Californian orange trees.

For some purposes it is important to count the minimum number of breeding animals in the course of a year. When this gets small, the frequency of a character in the population may change by mere chance, quite apart from natural selection. Dubinin, who first demonstrated this, called this change by a Russian phrase translated as the genetico-automatic process. I am afraid I may be accused of bourgeois prejudice, but I prefer the shorter term "drift" used by the American, Wright, who produced the theory of it, but actually observed it later than Dubinin. To measure drift we have not merely to catch thousands of animals, of one species, but to classify them according to their colour, shapes and so on, and to breed from at least some of them. And we have to repeat this in a number of seasons, to see if there is a steady evolutionary change, or only a random one. In fact counting wild animals is a whole time job.

5

Why I Admire Frogs

FROGS are beginning to spawn in the ditches of southern England. If we get a spell of frosty weather, as we well may, they and their eggs are going to suffer heavy casualties. I shall be sorry if they do, for I like frogs. Under the general term frogs I include all the tailless four-legged amphibians. In England we only have three native species, the common frog, the common toad, and the natterjack toad. But in other countries there are plenty of animals obviously related to our frogs and toads, but no closer to one than to the other.

The frogs in this broad sense are one of the three living orders of amphibians. The others are the tailed amphibians such as newts and salamanders, and the legless, tailless, and generally blind tropical amphibians which live underground and lead a life rather like our earthworm. Within the amphibians the frogs have developed in some ways as men have done among the mammals. They have lost their tails, their hind legs are a lot bigger than their front legs, and their heads are relatively large. Of course they have not developed their brains as we have, and though they use their hands for clasping, and sometimes for climbing, they are not organs of skilled work, as the beak of a nest-building bird is.

Our frogs are limited by the fact that their eggs must be laid in water, and they pass their first few months as tadpoles. This means that they can never go very far from stagnant or slowly flowing water. Others have got over the handicap in various ways. Many frogs in South American forests spend their whole life in trees, and lay their eggs in the water which accumulates between the stems and leaf-sheaths of plants which grow on the tree bark. In countries where there is daily rain in the breeding season eggs are often laid out of water. Several tropical frogs make nests, usually by sticking leaves together, in which a mass of eggs are placed.   Sometimes these are placed actually hanging over a stream or pool, so that as the tadpoles hatch, they drop into the water.   Others are placed so near to water that the tadpoles have not far to wriggle.

Quite a number of frogs burrow into the ground, including one species which is quite common in France.   In one Japanese species a pair burrow into the bank of a pond above water level, seal up the entrance to their hole, fill it with eggs, and then make a tunnel opening under the water through which they leave, and the tadpoles follow them on hatching. . In many cases where eggs are laid out of water the parents spend many hours kicking the egg mass with their hind legs until they have made enough froth to provide the air needed by the eggs during development. Other frogs carry the eggs about with them.   The male "midwife toad" which is found in France, carefully collects the string of eggs laid by his mate and carries them wrapped round his hind legs till they are ready to hatch.

Only one frog, a West African species, bears living young. This enables it to live on damp mountain sides at some distance from water.   But several species do something even odder. The female lays the eggs into a pouch covering her back, and here they develop, receiving nourishment from the mother like embryos in the womb of a mammal.

However the queerest habits of all are those of a Chilean frog, *Rhynoderma darwinii*.   Here the eggs are laid on the ground. Soon afterwards a number of males surround a mass of eggs, watch them for a fortnight or so till the tadpoles are beginning to wriggle, and then eat them, one by one,   However they do not pass into their stomachs.   Many male frogs have a bladder under the skin on each side of their neck, or rather where the neck would be if they had one.   This can be filled with air, and is used for croaking.   It is very small in our common frog, but fairly large in the edible frogs which exist in several parts of England, and are much noisier than the common ones.   In *Rhynoderma* the tadpoles swallowed by the males are passed into the croak sacs, which enlarge enormously.   When they hatch, the croak sacs produce a sticky secretion which plays the part of milk.   They remain there, growing till their legs have developed.

Finally the male, apparently with considerable difficulty, forces them out of his mouth, and they start life on land. I do not know how long it takes before he is able to use his sacs for croaking again. Perhaps the oddest feature in the whole situation is that the little frogs inside a male are not necessarily even his own children. It is queer enough, if you find a frog with young ones inside it, to discover that it is a male, still queerer to find that it may not be their father. Several fish species behave in a similar way. The male holds the eggs in his mouth, sometimes for several weeks, until the little fish can live an independent life.

I have not given anything like a complete account of the ways in which frogs have overcome the handicap of having to breed in water while living mainly on land. But I think I have given some idea of their variety.

Britain is very badly off for frogs and toads. We have only three native and one or two introduced species, as compared, for example, with fourteen species in France. The reason is that they find migration very hard. Not only are they killed by salt water, but a fairly dry range of hills, such as the chalk downs, is a barrier which they can hardly pass. A foreign species has recently colonized Romney Marshes, in Kent, but unless some human being helps them, it may be centuries before one crosses the hills and gets into another river basin.

I should like to see several foreign species established in England, notably the midwife toad and the spade-footed burrowing frog. They live in climates not unlike ours on the continent, and I have very little doubt that they could thrive in England or Scotland. But I doubt if it would be right to let them loose. When a foreign animal is introduced into a country it may die out. But if it does not, it often becomes a pest.

The reason is that it has been taken away from its natural enemies. For example in many parts of Europe storks do a great deal to keep the numbers of frogs down, and we have no storks. A foreign species may also wipe out a native one, either by direct competition, or by introducing diseases. This is one way in which the human races compete. For example measles, introduced by Europeans, killed off huge numbers of natives of the Pacific Islands. But yellow fever, to which West Africans are fairly resistant, has prevented Europeans from colonizing West Africa as they have colonized Brazil, which has a fairly similar climate.

We have no body of experts which undertakes the biological planning of Britain, comparable with the vast organization which is planning the afforestation of the South Russian steppes. Until we have an institution which will consider the effect of any introduction on our agriculture, fisheries, river conservation, bee-keeping, and so on, it is better not to add new animals to our rather small list. Nevertheless I should like to see some of the beautiful tree frogs introduced into warm valleys in Devon and Cornwall, where they would probably do well, and several larger frogs and toads would be a delightful addition to our countryside, provided they did not kill off native animals.

6

The Hungarian Invasion

NO, this is not another accusation against the Hungarian Government. It is an account of the successful occupation of a part of England by Hungarian frogs.

Until recently, the animals of England included two species of frog, and two of toad. The common frog, *Rana temporaria,* is found all over the country. The edible frog, *Rana esculenta,* is not so common, and has very likely been introduced, either by monks, epicures or amateur naturalists. It has a patchy distribution. The nearest colony to London stretches from Richmond Park across the Thames to Teddington, though the frogs are only common at a few points in this area. There used to be a colony in Kentish Town. All the other areas where it is found, so far as I know, are in south-eastern England. These frogs are rather larger and more striped than the ordinary kind, and the males have vocal sacs behind their jaws which bulge out to a surprising extent while croaking in the breeding season.

The common toad is found everywhere, but the natterjack toad has a patchier distribution. It is particularly fond of sand-hills near the coast, and one of the places where one can be sure of finding it is near Southport and Ainsdale in Lancashire. There is no suggestion that it was introduced by human beings.

Professor A. V. Hill, the physiologist, who was a Conservative M.P. in the last Parliament, used to import frogs of the species *Rana ridibunda* from Hungary. They are much larger even than the edible frog and some of them have a very bright green colour. He used these frogs, or rather their legs, for very accurate work on the physiology of nerve. The nerves of a warm-blooded animal die very rapidly after its death. Those of a frog continue to conduct for hours or even days after removal if they are kept moist. Among the things which Hill measured was the extremely small amount of heat produced when an impulse passes down a nerve fibre. As it would take several million nervous impulses to raise the temperature of a nerve by one degree, the measurements have to be pretty accurate.

About 1936 he gave several pairs of these frogs to Mr. E. P. Smith, also a Conservative M.P., and dramatist, who has a house with a pond near the Royal Military Canal, which runs between Romney Marshes and the rest of Kent. The Hungarian frogs bred in this pond, and after a few years some of them got as far as the canal. They spread along it in both directions, and during the war they occupied most of Romney Marshes.

They are not only larger than the common frog, but more aquatic. They spend most of their time in the water, and when they come out they usually stay near it and jump back when frightened. They seem to have killed off the ordinary frogs and toads in the marshes, or at least reduced their numbers considerably, whether by competition or more probably by eating their tadpoles.

Their most striking characteristic is their voice, which is extremely loud, at any rate in the breeding season. They are the form found in Greece, and which provided the chorus for Aristophanes' play "The Frogs", which is well known in Murray's English translation. According to Aristophanes they made two noises, which are transliterated, "Brekekekex" and "Koax, Koax." The Brekekekex sound, which they make during the day-time, has been compared with that of a machine gun, the Koax with the noise made by a cat when you tread on its tail. To some listeners the "song" sounds much more like Brekekekesh Koash. It is probable that the Athenians pronounced the letter Xi, which we render as X, much more like SH.

I suggest that the Classical Association, which is interested in how ancient Greek was actually pronounced, should send down a party to Appledore next May, when the frogs are breeding, to determine this point. Our knowledge on such doubtful matters rests partly on the transliteration of Sanskrit words. Sanskrit is handed down orally, and great care is exercised to pronounce it properly in religious rites. Frogs are probably even more conservative than Hindus in their pronunciation. Some of the Marshmen find the frogs' song objectionable, and

I should like to be able to use it as political propaganda, were it not that several naturalists who have been to the marshes think their complaints are exaggerated.

It is rather unlikely that these frogs will spread up the hills into the rest of Kent, and invade England. But they could colonize other marshy regions, and will do so if anyone transports them there. In 1939 Professor Hill dumped about fifty of these frogs into the Cam near Cambridge, but they seem to have died out. This is probably because when put in a river they scattered to such an extent that no male could find a female in the spring of 1940. If he had put them in an isolated pond some would probably have stayed and bred there, and they would have spread into other stagnant water and established themselves.

In captivity they will mate with the edible frogs and produce tadpoles, but no one knows whether or not the hybrids are fertile. Dr. Malcolm Smith, of the Natural History Museum, who recently described the Romney Marsh colony to the Zoological Society, hopes to test the fertility of the hybrids in a few years. At any rate *Rana ridibunda* and *Rana esculenta* inhabit different areas, and where these overlap, as in parts of France and Germany, they seem to keep fairly distinct. So it looks as if we had got a new English wild animal.

7

A Great Event

WHAT was the most important event in 1944? Some would say the landing in Normandy, the break-through at Avranches, or the liberation of Paris. Others would think that some of the great victories of the Red Army, in which more Germans were put out of action than in France, were even more important.

There is, however, a third point of view. According to the London Bird Report for 1944 "The outstanding event of the year was undoubtedly the nesting of the Little Ringed Plover at a gravel-pit in Middlesex." Only once before has this species been found breeding in Britain, and it has not been seen near London since 1864. I confess I do not take the colonization of this country by birds quite so seriously as does the editor of the *London Naturalist,* but if human society had developed a little differently this event might have been of the highest political importance.

For among many primitive people rare birds are of great economic value. Wallace's account, in his *Travels on the Amazon and Rio Negro,* of the Uaupe Indians in Columbia is a classic. They kept large numbers of birds, including parrots which were allowed to fly about freely, but returned to be fed, egrets, and eagles which were kept in large houses, and ate two fowls a day. They bred these birds for the sake of their feathers, which were used for men's head-dresses. The parrots' feathers were not of the natural colour, but prepared by pulling out a green feather and then rubbing in the secretion from a toad's skin. The new feathers which grew after this treatment were yellow.

A well-dressed middle aged man also wore cords of monkey's hair down his back, a bead necklace, a large white cylindrical stone, a belt of jaguar's teeth, a pair of garters, and rattles on his ankles. The ladies wore garters only. A powerful chief was said to be the possessor of great wealth in the form of feathers and jaguar teeth, which he had won in a series of wars, but would not show to white men, for fear of losing them. These wars, wrote Wallace "are fought for the sake of their weapons and ornaments, and for revenge of any injury, real or imaginary." Wallace wrote in 1853, and had certainly not read Marx, but he made a very sound analysis of the economic causes of war.

In our own culture we do not greatly value animal products, though some furs are fairly expensive. But quite small quantities of a yellow metal, gold, and a sparkling stone, diamond, are valued at as much as a lifetime of hard work. We fight wars, for example the Boer War, for such objects, and when we have got them we shut them up in underground vaults. If we valued rare bird feathers as we do gold, the appearance of the Little Ringed Plover in England might have precipitated a Feather Rush, like Charlie Chaplin's Gold Rush. Or if it had happened a little earlier it might have decided Hitler to invade Britain rather than the Soviet Union. Actually this bird is common enough in Europe, but its introduction into a new country might well have been a menace to monopolists on the continent.

From the strictly scientific point of view the interest of such observations as this is rather slight. Indeed the fact that a rare bird is seen in England is more important to continental than to British zoologists. Was there an excessive number of these birds in France in 1944? Or did their food run short? Was their advent connected with the much larger invasion of Waxwings which took place in the same year? Such are some of the questions which a zoologist might ask, and whose answer would help us to understand the laws governing animal populations.

On the whole I am glad that rare animals do not have a great commercial value. If so their habitats would be kept secret, as are the details of the geology of oilfields, and the study of geographical distribution would be even harder than it is now. Certainly some of the habitats are unexpected. Thus *Pseudosinella religiosa,* a microscopic insect belonging to the order *Collembola,* has only been found in the crypt of Lund Cathedral in Sweden. This crypt, by the way, is interesting to entomologists for another reason. It is embellished with the carving of a louse about a foot long, sucking the blood of a lamb of the same size. The louse represents the State, and the lamb the Church, though some people have taken the parable the other way round.

But the habitat of one particularly rare and interesting animal is absolutely unknown. The late Major-General Hearsay (he really was called that) made a collection of insects which passed to the British Museum on his death. While fording a river during a battle in India he was observed to catch several specimens which he pinned to his pith helmet and catalogued later. Most of his trophies were quite common, and of no value, as he rarely stated where he got them. But one, an insect called *Prophalangopsis obscura,* is of extraordinary interest. Not only is it the only known member of its species, but the species is so unlike all other insects that it has been put in a family of its own. And it combines so many primitive characters that it is regarded as closely resembling the common ancestor of the crickets, the long-horned grasshoppers, and several other groups. Unfortunately the major-general's service was so varied that we do not know even in which continent he found it.

But my favourite story is of the Peruvian gentleman who was disturbed by a noise in his garden one night in the middle of Lima. He fired his revolver at it, and next morning discovered the corpse of a large animal unknown to science, something like a guinea-pig with a bushy tail, which was called *Dinomys,* and also assigned to a new family. A number more of these animals have since been found. We are not, however, told whether this was the most important event of its year in Peruvian history.

VIII

HOW ANIMALS BEHAVE

I

The Robin

MOST original scientific work is published in special journals, such as the *Proceedings of the Royal Society*, the *Journal of Physiology*, the *Observatory*, or the *Philosophical Magazine*. Occasionally it is published in book form, but most scientific books are summaries of work by many authors, which has already been published in periodicals.

The vast majority of papers or books containing original research are quite unreadable by the general public, because they assume so much special knowledge. A bright exception is David Lack's *The Life of the Robin*,* which describes the author's researches on a subject with which everyone is familiar. It should be in every school and public library not only because of its intrinsic interest, but because it gives an easily followed account of some scientific research, a human activity which, whether or not we understand it, is of great social importance. Lack studied the robins in about 20 acres of Devonshire. The first thing to do was to catch them all, and let them out again with coloured rings on their legs for identification. His area held from 11 to 19 adult robins.

Why does a robin sing? The song has been interpreted as courting, and as a sheer expression of joy. Actually both the song and the display of the red breast seems to be a substitute for fighting. One of the fundamental facts in robin life, as in that of many other birds, is territory.

In autumn every male robin occupies a territory of about an acre, in which he spends most of his time, though he occasionally feeds outside it. Within this area he sings and struts vigorously. If another male appears in it he sings at him violently, and may peck him. Usually the intruder goes off; occasionally he may fight the quarrel out, or sing it out, and occupy all or part of his

* F. H. & C. Witherby Ltd., London, 7/6.

rival's territory.   About the new year females arrive in the male's
territory, and may or may not stay there.   There is a great deal
of singing, which may serve to attract the females, but the
females have an absolutely free choice of mates.   One of the
visiting females settles down with a male in his territory.

After marriage the female helps the male to defend their joint
territory.   But they do not mate till March or April when the
nest is at least partly built.   At this time the male also brings his
wife presents of food, and later helps to feed his children.   Occasionally a female leaves her husband for another male, and two
cases of bigamy have been reported; but as a general rule a
couple keep together till the young have left the nest.   However
during the late summer and autumn the sexes separate, and the
hen usually finds a new mate next winter.   The males generally
stay in one place for life, but most of the females migrate, some
of them passing the winter in France.   In fact the male robin is
the home-lover, and the female wanders about and chooses her
mate.

Why do robins behave as they do?   Lack did a number of most
interesting experiments with stuffed birds.   Both males and
females will threaten stuffed specimens placed in their territory,
and may attack them.   If they cannot destroy the "enemy" they
will desert their nest if it does not contain young.   What is more
remarkable, they rarely attack a stuffed robin whose breast has
been dyed brown, but will threaten a small bundle of red breast
feathers.   It is useless to say that a robin has poor eyesight.   They
can recognize their own mates.   Such conduct certainly seems
irrational to us.   But if an intelligent robin discovered that human
beings spend immense amounts of energy and thousands of lives
in digging a yellow metal out of a series of holes in South Africa
to bury it in another hole in the United States, they might not
think our species very rational.

"Who killed Cock Robin?" is quite an important question
when posed scientifically.   Only by answering such questions
can we determine whether natural selection is really the driving
force behind evolution, as Darwin thought.   A robin can live
for as long as eleven years.   A pair can also hatch out two broods
of five or more eggs a year.   Now the population of robins is ·

fairly steady. If it increased by only one tenth per year, it would grow to 13,781 times its original number in a century. An average pair of robins which live to be adult probably produce at least twenty eggs. And if they lived under very sheltered conditions they could probably produce 100. Actually only two eggs would be needed if there were no untimely deaths. So we may say that the struggle for life accounts for nine out of ten robins actually begotten, and forty-nine out of fifty which could be begotten.

Lack concludes that where cats and boys are common, about half the nests are destroyed, though most of the nests at Dartington were successful. A few young die in the egg or in the nest, but about three quarters of those which leave the nest are dead within a year. Even of those which survive the perils of youth, more than half die within the next year. The average life of a robin seems to be about one year or less. So far as the causes of death are known, cats, mousetraps, and motor vehicles head the list, but probably cold and starvation are equally important. A robin redbreast in a cage may, as Blake said, put all heaven in a rage; but, if like Mr. Lack's cage, it is thirty feet long, twelve wide, and six high, the robin is quite willing to breed in it, and is much safer than outside.

Lack is, I believe, a schoolmaster,* and has done all this work in his spare time. I want to see a social organization in which everyone will have the leisure for scientific work. There is any amount to be done, even in towns. No one knows half as much about the sparrow as Lack has discovered about the robin. There are huge gaps in our knowledge even about cats. We know still less about the habits of less conspicuous animals. As the pressure of war work falls off, I hope that many people who have been much too busy since 1939 will realize that in work of this character they can not only find intense enjoyment, but make real contributions to science.

* I am glad to say that he is now head of an ornithological institute.

2

The Starling

OUR migrant birds from the south are just (March, 1948) arriving or have recently arrived to spend the summer in Britain. Some have not come very far, but a good many of the swallows have flown all the way from South Africa. And the birds which have wintered in Britain are going northwards and eastwards. It is of one of these species that I am going to write, namely the starling. It is particularly interesting because it may give an answer to the question of how a species originates. For the starling seems to be splitting up into two species, as Darwin believed that species sometimes do. The starling is a remarkable bird in a number of ways. It is much more social than most British birds. During most of the year many starlings roost in huge communal roosts with up to a hundred thousand members, from which they may fly for twenty miles each day to feed. Whereas the robin, to take a familiar example, spends most of its time in quite a small area, and quarrels violently with other robins except its mate. Again there are about twice as many males as females, so it is not uncommon for a female starling to have two husbands. This again suggests that they are not so quarrelsome as most of our small birds.

But the most remarkable fact, first clearly proved by Dr. Bullough, of Leeds University, is that while the starlings which spend the summer, and breed, in England, spend their lives here; those which come here about October and leave in March, breed in Europe. The Scottish winter visitors go to Norway. The English winter visitors go as far east as Sweden and Latvia, and probably into Russia. This migration had been known for some time, but Dr. Bullough was the first to distinguish the two types by the fact that the British, or sedentary, race, has much lighter coloured beaks, especially in December and January. This was confirmed by putting coloured rings on the legs of a

number of starlings, and observing that the dark beaked ones
went away in March.

The reason for the different behaviour of the two types is this.
During the winter the continental birds have very small ovaries
and testicles, and show no sexual behaviour. But although these
organs get smaller in the British race after the breeding season in
April, they increase in size in autumn, and there is a good deal of
love-making, and sometimes even mating.

Now it seems to be a fairly general rule that the presence of
sex hormones not merely makes a bird amorous of the other sex,
but makes it attached to its place of breeding. The continental
starlings fly back to their breeding places where their sexual
organs mature in spring. The British birds not only never
leave Britain, but they never move permanently away from
their homes except in their first year of life, before they are
sexually mature.

It is not safe to guess too much about the emotions of animals.
But it certainly looks as if the emotions of birds towards their
homes and their mates were similar. This is true of some human
beings too. I have known at least one man who complained
that his wife seemed to think she was married to her house, not
her husband.

There seems to be little doubt that membership of the two
"races" of starling is hereditary, and fixed for life. This means
that they do not breed together, any more than the sewer rat and
the ship rat. Very possibly an occasional abnormal continental
starling remains in Britain, or a British one goes east. It is
probable that if they got the chance, they could and would
breed together, but this is not known.

The two races differ in a great many habits as well as their
migration. For example, the continental starlings never seem
to roost in towns, whereas many of the British ones do so.
Among the well-known London roosts are the Marble Arch,
and St. Martin's, Trafalgar Square. Again the British birds are
rather less social than the continental ones. Some never seem to
join communal roosts. Others only do so in July and August
after the breeding season. But there is no hostility between the
two races. On the contrary, the British starlings are more

tolerant of foreign birds than of their compatriots, probably because the latter are more apt to be sexually active, and thus arouse their jealousy.   There are slight differences in the feathers of the two races, and in their eye colours, but not enough to enable one to distinguish between them with certainty.

It is very unlikely that the separation goes back even as far as ten thousand years.   For ten thousand years ago, as we know from plant remains, England was a good deal colder than it is now, and would not have been a suitable winter refuge for birds.   From what we know of the rate at which species originate, it is likely to be another twenty thousand years or more before the two races of starling evolve into a pair of species.   One can guess at some of the possible lines of divergence.   The continental starlings have farther to fly, and may be expected to develop rather more powerful wings.   The English ones spend more of their times in holes and therefore rub their feathers more. They are likely to develop stouter breast feathers.

Other species of birds are probably splitting in a similar way. Thus Promptov finds that the chaffinches of Ukraine are divided into two groups which do not inter-breed, on the basis of their song.   Naturally we cannot expect to see species formed under our eyes in nature, though something very like it can be achieved in the laboratory.   Thus Kozhevnikov produced a race of the fly *Drosophila melanogaster* which breeds true but gives sterile eggs when mated with the parent race.   Nevertheless our starlings give us a fairly good idea of one of the many ways in which a species can split in two.   I wish we could resurrect Darwin, if only for five minutes, to tell him about it.

3

Clever Birds

WHEN biologists accepted Darwin's theory that men were descended from animals, they naturally tried to find some way of estimating intelligence in animals.

The chimpanzee and orang-outan came out very high. In particular Köhler found that some chimpanzees showed a good deal of intelligence in the use of solid objects as tools. They would pile boxes on one another to reach a suspended fruit. If given two canes, neither long enough to reach a fruit outside their cage, they would fit the narrower one into the broader in order to make a rod which was long enough. Clearly chimpanzees have at least a rudiment of the sort of intelligence which is developed in a skilled craftsman.

On the other hand Yerkes tried in vain to find in them the germs of mathematical intelligence such as we display when we take the second turn on the right, or bring home three loaves, not two or four. One of his methods was to give the animals a choice of a number of boxes, say nine in a row, of which some were open and some shut. If food were put in the left-hand box of those which were open, the apes sometimes got the idea. But they never managed to grasp the idea that food would be found in the leftmost open box but one. It was of course necessary not always to put the food in the same box. Animals easily pick up clues from small marks on wood, smells, and so on.

Other mammals were not much better, though two pigs did surprisingly well. Coburn and Yerkes managed to teach two crows to choose the farthest open door on the right, but it was a slow business, and they never mastered the choice of the first from the left completely. On the other hand Sadovinkova reported much greater success with two canaries, a goldfinch and a bullfinch. One of the canaries was a dunce, but the others all beat Yerkes' apes. The goldfinch, after two lessons, always chose the second, starting from the left, of four open doors. When anything from three to nine doors were left open, he made one mistake. Later on he learned to pick the middle door out of three, five, seven or nine open ones.

Sadovinkova's work was not followed up till 1943, when a Finnish biologist, von Haartman, repeated it using fly-catchers, a lark and two bullfinches. He reports his results in the first number of *Behaviour*, a journal published in Holland, and devoted to animal psychology. Other authors describe the behaviour of wasps, mice and sticklebacks in the same journal. The young of a pair of flycatchers were put in a box near the old nest, and the parents had to learn to visit the first open door below the top one of a series. The mother finally managed to solve the problem correctly ten times running. The father generally made three or four mistakes out of ten.

The larks and bullfinches were kept in cages with a range of boxes from which to choose food. The bullfinch were "punished" for a wrong choice by letting down a screen in front of all the openings. The lark was too easily frightened for this to be done. The hen bullfinch was the best learner. She could choose the second opening from the right out of a variable number, and when the distance between holes varied from two to six inches.

It is easy to see why small birds should have this very specialized kind of intelligence. They have little or no sense of smell, and owing to the hardness of their beaks they can have little detailed sense of touch. They rely mainly on sight and hearing, though of course they may have a sense unknown to us which helps them in migration. They find their way through woods and thickets, or return to a nest in the middle of a meadow, by sight. So not only is their sight keen, but they can recognize objects and patterns to a surprising extent. The fact that they can learn tunes proves that they have an equally good memory for sound patterns.

They are remarkably good at building nests with such a clumsy organ as the beak, but we could not expect them to develop the manual skill and insight of an animal like the chimpanzee with a pair of hands. Only one bird is known to use a tool. This is a finch in the Galapagos Islands which uses a cactus thorn for picking out insects from under bark, as other birds use their sharp beaks. It would be interesting to test its intelligence in other respects. Our own thrush, of course, uses hard stones against which to crack snail shells, so it is near to the tool-using stage, but I hardly think it can be regarded as a tool-user.

These experiments are interesting in two ways. They show that small birds have a good deal of intelligence, as many people who have kept canaries and budgerigars have said for a long time. And they suggest that the kind of intelligence on which psychologists tried to concentrate is not the most important kind. It is a capacity for picking out resemblances and differences rather than one for solving the problems which arise in real life.

It is of course much easier to devise tests of this abstract kind, either for birds or men, than tests which will measure one's capacity for getting a machine or a committee to work properly, still more for inventing new kinds of machine or of social organization. But I suspect that the mental qualities which psychologists can so far measure, or at least grade, are not the most important ones for human society. There is nothing surprising in this. They have only been at work on these problems for a generation or two.

It was a long time before physicists could measure the qualities of a metal or a textile material as well as experienced craftsmen could do. Psychologists have already made great steps in the right direction, but they still have a long way to go. And they do not always take account of the fact that the qualities which make for success in different kinds of society are very different. Men who achieve peerages in England would often be jailed in the Soviet Union, and men who would have been in high positions there have been jailed here. Probably both these types of successful men would have been failures in the Middle Ages or in many primitive human societies. In fact human abilities can only be judged against a social background.

So perhaps it is a good thing that some psychologists should work on animals, where this kind of bias does not operate. It may help them to think more clearly about men and women.

4

How Bees Communicate

EIGHT years ago I gave an account in the *Daily Worker* of the early work of von Frisch and others on the language of bees. In July 1947 I was at the London Zoo with Professor Hadorn of Zurich. We watched bees coming in to the glass-fronted hive laden with pollen of different colours in the bags on their legs. He was able, by watching them, to tell me from what direction they had come, and roughly from what distance. So will you be, after reading this article. The facts previously known were these. When a bee has found a rich source of pollen or honey she comes back to the hive, and before handing it over to the other workers whose job it is to store it in the comb, she does a peculiar "dance". During the dance other bees touch her with their antennae, so that they know what smell is associated with the kind of food in question. They then fly off to visit flowers of the same kind, or dishes of sugar water impregnated with the same smell, for example of peppermint.

When the flower or the sugar-water is placed within fifty yards of the hive, bees fly out in all directions to visit flowers or dishes with the same smell. But when they are placed at distances over about a hundred yards, they not merely fly out in the right direction, but for the right distance. As some of them arrive before the original finder has unloaded her honey or pollen it is clear that she must have told them in which direction to go, and how far.

Von Frisch has discovered how the information is conveyed. If the food is within fifty yards the finder always dances round and round. If it is more than a hundred the dance is quite different. She goes forward in a certain direction for an inch or two, waggling her abdomen, then runs back without any "dancing" and repeats the dance again and again. The more she has found, and the sweeter the sugar-water, the longer the

196

dance lasts, and therefore the more other bees are able to learn what smell is associated with food, and the more go to look for it.

If the finder dances the round dance they go out in all directions, but not for further than a hundred yards. The other kind of dance gives them the direction. The dances are generally carried out on the comb, but sometimes on the landing stage in front of a hive. If it is horizontal, the dancer moves in the direction of the food, and the other bees fly out in the direction of her dance.

If however the surface of the comb is vertical, something much odder happens. As the day goes on, the dancer moves in different directions after coming from the same place. Supposing the food is south-west from the hive, then at 9 a.m. the dancer moves horizontally to the left, at noon she moves at forty-five degrees upwards, at 3 p.m. vertically upwards, and so on. In fact a dance upwards means that the food is in the same direction as the sun, a dance to the right that it is to the right of the sun, and so on. It is most remarkable that bees know the direction of the sun, even in cloudy weather. The distance is given by the rhythm of the dance. Food only 150 yards away elicits a dance with 40 tailwags a minute. This number sinks to 20 when the food is half a mile away, and to only 8 at a distance of two miles.

Von Frisch believes that the same language is used by scouts which go out from a swarm of bees and come back to tell it where they have found a place suitable for a new hive. But this is uncertain, for swarming is rare, whereas hundreds of observations can be made every summer day in an ordinary hive.

Besides the dances the bees have at least one other "word", namely a sweet smell which they make when they have found rich food, and which attracts other bees to the place.

These observations seem to have a great philosophical importance. It is often said that animal "language" is a mere expression of the emotions, and cannot convey statements of fact. But it is clear that the bees can tell each other not merely that they have found food, but where they have found it. It is true that the bees' language seems to be inborn, and not learned like ours. It is like that of the young lady in Shaw's *Back to Methusaleh* who emerges from an egg talking perfect Shavian English. However some birds have to learn a good deal of their language. It is also clear that bees have an amazing sense of direction. If a hive is turned round the dancer moves over the comb in a curved path as if she had a compass needle in her head. Perhaps she has some kind of magnetic sense which we lack. Her perception of rhythm must also be superior to our own.

A reader may well ask whether it is not possible that von Frisch is pulling our legs, or at least letting his imagination run away with him, and has invented the amazing story. The answer is that although he made some mistakes in his interpretations of dancing, most of his earlier work has not only been confirmed, but applied in practice by Gubin, Komarov and others in the Soviet Union, as well as by von Frisch himself in Germany and Austria.

Red clover is normally fertilized by bumble bees, and does not set seed without fertilization. Bumble bees are not common enough to fertilize an area of an acre or more of red clover. And ordinary bees prefer other flowers as their probosces are not long enough to get all the nectar of a red clover flower. The following method is therefore used. Beehives are brought near to the clover fields. Glasses containing sugar water and red clover blossoms are placed among the clover. Bees soon find them and come back to dance. Their comrades fall for the propaganda and search for flowers with the correct smell. A few of them find the sugar-water. The majority search the clover flowers. They do not find much nectar there, but in their searches they carry pollen from one flower to another. Enough of them find sugar-water to keep up the stream of propaganda in the hive.

The system rather reminds me of the football pools where a few people win large prizes, but the vast majority merely enrich the organizers of the pools and keep the postal workers busy. Economically it pays the seedsmen. For an expenditure of about twelve pounds of sugar per acre over five weeks von Frisch got an increased yield of 36 pounds of clover seed per acre. As a pound of clover seed cost as much as 16 pounds of sugar this was a good bargain, except perhaps for the bees.

Possibly an even bigger return could be got by inducing

bees to visit orchards. It is important that the bees visiting prepared sugar-water should if possible perch on fruit blossom and also suck up juice into which blossom has been crushed to give it the right scent.   I do not know whether we shall be able to learn the language of ants, and get them to clean our kitchen floors instead of raiding our sugar.   But I am quite sure that research on these subjects will tell us things which we need to know, not only about animals, but about human societies.

5

How Bees find their Way

I HAVE written several articles on the work of von Frisch on the senses of bees, and their methods of communication. Quite recently he has made a discovery which may clear up a lot of the so-called mysteries of animal behaviour.

The bees certainly perceive forms and colours. More accurately they can be trained to distinguish them. Their colour sense is excellent, but their form sense is nothing like as good as our own or a bird's. But they also perceive a quality in light which we do not, namely polarization. When light passes through certain crystals it is of course bent out of its path. But instead of one ray coming out of the crystal two do so with rather different properties.

The electrical vibrations which constitute ordinary light are in all directions at right angles to its direction of travel. But in a beam of polarized light coming out of a crystal they are only in one direction. For example if the light is going northwards the vibrations could be up and down, or east and west, but always in the same plane. One can make a crystal prism which only lets through light polarized in a particular plane, and by turning it round see in which plane a particular beam is polarized. Sugars, and many other substances with asymmetrical molecules, rotate the plane of polarized light passed through water in which they are dissolved, and this property is used to measure sugar with great accuracy.

Von Frisch found that bees returning from a good source of honey or pollen indicate its direction to their comrades by a peculiar dance. If this dance is done on a flat surface it consists of a series of runs in the direction of the food, with more leisurely returns in a curved path. If it is done on a vertical surface, such as the honey comb, upwards indicates the direction of the sun. So if the sun is in the south and the food to the south-west, they run upwards at an angle of forty-five degrees to the right of the vertical. Von Frisch found that if he gave them a flat surface to dance on, and put over this a double glass sheet with crystals between the glasses which only lets through light polarized in one direction, the direction of the dance was altered.

What does this mean? Direct sunlight is not polarized, but the light from the rest of the sky, particularly from clear blue sky, is so. So is light reflected from anything, though it is not so strongly polarized as light passed through some crystals. That is why motorists can cut out the glare of reflections from a wet road by using "polaroid" spectacles or windscreens. The bee is guided not so much by the sun, which may be hidden by clouds, as by the polarization of the light from the sky. And it is fooled by substituting artificially polarized light for the naturally polarized daylight.

We distinguish the up and down direction without thinking. The bees, at any rate when there is even a small patch of blue sky, can distinguish the direction of the sun, even when the sun is behind a cloud. Of course we have no idea what polarized light looks like to them, but nor have we any idea whether their colour sense is at all like ours. It may be more like our sense of musical pitch or of smell. But the result is as if what to us is a uniformly coloured surface were ruled with fine lines in one direction, gradually changing in the course of a day.

A great many other insects are remarkably good at finding their way back to their nests, and zoologists will probably be occupied for some years to come in finding out whether they too perceive polarization. It is possible, but unlikely, that birds have similar powers. It is unlikely because migrating birds fly over the sea at night, and keep a pretty true course.

A number of suggestions have been made as to how birds find their way, and many have been disproved. They might, for example, be sensitive to the earth's magnetism, like a compass. But if so they would be upset by strong magnets, which they are not. But the discovery concerning bees suggests that they may be aware of some directional quality in their surroundings to which men are insensitive. This may turn out to be something well known, like polarized light. It may also be that they rely on some happening which physicists have not detected. If so the discovery of how birds are guided during migration will also be the discovery of a new physical phenomenon, perhaps of very great importance.

The interest of von Frisch's discovery is two-fold. It is the first time that a sense has been found in animals quite different from any of our own. No doubt a dog can smell much better than we, and his sense of the direction of a sound is also much better than our own. But we know what smell is, and can roughly locate a sound. If a dog were aware of magnetism or of radio waves, we should have something more like the bees' new sense.

Secondly, it warns us how careful we must be in interpreting the behaviour of animals as if the world appeared to them as it does to us. We are apt to credit them with wonderful instincts or intuition when they merely perceive things which we can only detect with complicated apparatus. There is only one world, but it must appear very different indeed to different kinds of animals, and we may yet learn a great deal about it by studying animal behaviour.

6

Movies for Toads

PRIMITIVE men take it for granted that animals can think, and according to many religions they have souls which are judged after their deaths. Christian philosophers have usually denied that any other animals were capable of reasoning or had any rights. So, till the nineteenth century at any rate, animals were better treated in India than in Europe. Darwin took the view that the higher animals possessed most of the human faculties, though many of them poorly developed. On the other hand some of the more mechanistic biologists try to explain all animal behaviour on mechanical lines.

It is extraordinarily difficult to be sure that animals are thinking, and not picking up clues given by the experimenter. For example about 1910 a German at Elberfeld had some horses which could do fairly elaborate arithmetic, such as extracting cube roots. A French journalist, zealous for his country's honour, produced the story of a cat at Bordeaux which corrected the children's homework, mewing when there was a mistake in a sum and purring when it was right. However no one ever saw this cat, and plenty of people saw the horses. When a sum was put up on the blackboard, they tapped out the answer with their hooves. But they did not do it unless the teacher was there, and the psychologists who examined them finally concluded that they watched him carefully, and stopped tapping when he wished them to. Perhaps he made some signal deliberately; more probably they noticed slight changes in his expression or breathing.

A better idea of an animal's capacity for grasping numbers comes from the experiments of the late Dr. Honigmann on hens. He put hens in a cage with a narrow gap in the floor. Under this gap a board moved on rollers, carrying a row of wheat grains, of which the hen could pick one at a time. He glued down every second wheat grain, so that the hen could not remove it. After a while some hens learned only to pick at alternate grains. But they failed completely to conform to the situation when only every third wheat grain was free. If we like to put it that way, we can say that a hen can count up to two, but not to three. Other birds seem to be able to count up to five or so, or at least to notice a difference between four and five eggs.

Another of Dr. Honigmann's experiments was probably still more interesting. It has often been stated that animals cannot appreciate pictures. Certainly they recognize gramophone records. And female crickets will come to a telephone if a male of their species is chirping at the other end. But a dog rarely, if ever, shows any interest in a picture or photograph of his master or another dog. Nor is he interested in cinema films.*

The first animals which undoubtedly reacted to a moving picture are the common toad and the natterjack toad. They have very simple minds, if mind is the right word. They will only eat moving objects. The response to food is very characteristic. Although it is slow in its ordinary movements, a toad can flick out its tongue with very great speed and accuracy, and bring back a small insect or other food into its open mouth. It then swallows it if the taste is suitable. They are not interested in dead mealworms if they stay still. But a toad will flick and snap at a dead mealworm if it is dragged past the toad, especially if there is a well-marked background behind it. The toad snaps in the same way if the mealworm stays still, and the toad and the background are dragged past it. Toads will also snap at a film of a moving mealworm or other small animal. It might be argued that they react mechanically to any small object which moves or seems to be moving. But this is not the whole story.

If two toads are competing for the same food, and one of them gets it, the disappointed toad, especially if hungry, may flick its tongue at the eye of its successful rival. This is particularly common if the successful toad is the smaller of the two. Honigmann made films of toads eating, and showed them to

---

* Several correspondents claim that their dogs enjoy films, particularly of the "Wild West" type.

other hungry toads. I have myself seen them flicking the film star in the eye with their tongues. But this is most constantly done if the film is shown a little less than life size, and rarely happens if it is enlarged. The toad is not a noble animal. A toad will attack a picture of itself feeding as readily as that of another toad, but it is quite uninterested in films of other small animals, such as hamsters, feeding.

Encouraged by his success with toads, Honigman went on to show his films of moving worms to dragonfly larvae. These live under water, and shoot out their jaws at their prey. The film had to be shown projected onto a sheet of white paper pasted on the side of a glass tank, and the insects snapped at the pictures.

It is curious that so far moving pictures have only interested animals with very simple minds, such as toads and dragonfly larvae, and those with very complicated ones, namely ourselves. I think the reason is clear. A dog or a cat has mind enough to know that the moving picture is not a real happening, though a few dogs will watch films representing vigorous action. Men alone have negated this negation. We know that the hero is not really in danger of the electric chair, and that we shall not meet the heroine even if we stay at the stage door all night. But in spite of this, we manage to work up the appropriate emotions in a minor key, so to say.

I have described these particular experiments rather than hundreds of others which are constantly being made on animal behaviour, largely because I liked Dr. Honigmann, a refugee who worked at the London Zoo before the war, was interned, and carried out his work on toads in Glasgow until his death. I liked him partly because he so obviously liked toads. While one can go too far in treating animals like human beings, I believe that one achieves the best results, whether with animals, plants, or men, if one likes them as well as being interested in them.

IX

EVOLUTION

The Struggle for Life

ALL the biologists known to me accept the theory of evolution in some form, though a few postulate supernatural intervention at certain stages. As to why it happened there is far less agreement, but I think a big majority believe that natural selection is one of its main agencies. It is extremely hard to watch natural selection at work. To do so we must first show that some character is inherited, and then prove that animals or plants possessing it are more likely to survive to maturity or more fertile when they become mature, than the rest of the species, and that in consequence the character is spreading.

Probably the best demonstration of natural selection at work is that of Dubinin and his colleagues. He showed that within a species of fly, a particular type increased during warm weather, and decreased in the winter. He also showed that the kind favoured by warmth was commoner in towns than in the country, but no commoner on the sites of towns which had been destroyed by the Germans. By laboratory experiments he showed that the type which increases during the warm weather is much more easily killed by cold than the other type.

However, a really thorough study will require counts of the numbers of an animal species over many years, including a study of the reasons why members of it die at different periods. A study of an animal population on these lines has just been published by G. C. Varley. It has not yet got to the stage where natural selection was observed, but it gives an idea of how such work will have to be carried out in future. Clearly it is easiest to work with an animal which spends most of its life in the same place. He chose a gall fly which lays its eggs in the flowers of the common knap-weed (something like a large daisy), and spends most of its life as a grub inside the gall which it forms there.

Even when adult it does not generally move far. He marked

108 flies with spots of paint, and even after a fortnight found
none more than 22 yards away from the spot where they were
let out. From counts of flies and grubs, he calculated that a
female laid about 70 eggs in July 1935, 52 in 1936, and 200 in 1937.
Since there are about equal numbers of males and females, and the
species is not increasing or diminishing greatly in numbers, all
but two of these must die before maturity in an average year.

A few eggs were addled. A fair number of grubs died before
forming galls, but the biggest cause of mortality in summer and
autumn was from other insects which lay their eggs in or near
the grubs. When these eggs hatch, the gall-fly grubs are eaten
alive. The large majority of the gall-fly grubs each year die in
this slow and perhaps painful manner. Others were killed when
caterpillars ate the flower heads in which they lived. About a
quarter survived till winter, when most of the flower heads
with galls fell off. A large number of these were eaten by mice,
and others died when the ground was flooded. By spring 1938
only about one in forty of the generation survived. Parasites
and birds further reduced this number, and by July an average
of about two flies emerged from each batch of 200 eggs.

Varley went on to make similar calculations for the most
important parasites. The fact that twenty-four insect species
were found in the galls shows how immensely complicated even
a very simple community proves to be. However, the most
important parasite turned out to be dependent on the density of
the hosts. Each female of the parasite species searches about
100 flowers for gall midge grubs. In a year where they were
frequent she laid about 60 eggs, in a year when they were rare
only about 9. In fact when there are few gall midges, the number
of parasites diminishes sharply. When there are many, it
increases. So the parasites prevent the hosts from becoming
very numerous, but cannot destroy them altogether.

What is much more surprising, if birds or mice eat some of the
gall midge grubs, along with their parasites, this is a definite
advantage to the gall midge. It gains more from the
destruction of parasites than from the destruction of its own
species, and its numbers increase. This had been demonstrated
on theoretical grounds by Volterra in Italy, and by Nicholson

and Bailey in Australia, but Varley was the first to prove it by actual figures.

There is very little direct competition between the gall midges. When they are numerous there is a little competition for food, but when they are rare some females do not find mates, and lay sterile eggs, and these two effects roughly balance one another. The danger from overcrowding comes not from direct competition, but from increase in the number of parasites. Man was in the same condition till very recently. The death rate from disease in towns was so great that human populations could not increase beyond a moderate density.

By giving towns a water supply uncontaminated with sewage, and by other hygienic measures, we have put an end to this state of affairs. But unfortunately in other respects we are still in a state rather like that of gall-fly grubs. It is only public misfortunes which enable us to cope with human parasitism. As a result of the war we had to adopt rationing, and in consequence about a third of our people have better food and clothing than they had in 1939. Whether they will continue to get them much longer is another question. The peoples of Eastern Europe have only been able to shake off the worst of their human parasites after a ghastly period of oppression and war. The Communists of Western Europe, who would like to do the same thing without massacres and foreign invasions, are accused of trying to produce the chaos which they want to prevent.

The gall midges cannot think. They have to rely on birds to eat their parasites, even if a great many of themselves are eaten along with them. Men can think, but most of them find the process pretty unpleasant. It is time we did a little serious thinking along the lines of *Britain's Plan for Prosperity*. If we do not do so, we may find ourselves in a position rather too like that of the gall midges.

2

Man's Ancestry

WHEN Thomas Huxley first produced serious evidence that man was descended from an ape-like ancestor, his critics quite rightly pointed out that there was a large gap between men and apes, and that no fossils were known which bridged this gap.

For such hypothetical fossils they coined the phrase "missing links". Since that day two essential links have been added. Dubois found skulls and other bones in Java which he assigned to the genus *Pithecanthropus*. There was doubt as to whether they were apes or men. There is not much doubt now, because very similar skulls from the neighbourhood of Pekin were associated with stone tools, and palaeontologists are agreed that an ape-like creature which used tools deserves the right to be called man. So far as I know the first person to make this point was Marx's colleague Engels in an essay reprinted in *Dialectics of Nature*. Very likely the idea came from Marx. At any rate it is generally accepted now by people who would be horrified to be called Marxists.

These very primitive men had brains a good deal smaller than ours, prominent ridges above the eye sockets, no chins, and various other rather ape-like features. But they obviously used their hands as we do, though probably not so skilfully. Gigantic forms closely related to the Java and Pekin men have also been found in Java and China. We do not yet know enough about them to say whether they are likely ancestors of humanity.

In the last twenty years a new, and perhaps more important, link has been found in South Africa. Dart found the skull of a baby and since then Broom has found skulls and various other bones of a number of adults, in hard stalagmitic deposits in caves. These skulls are far more ape-like than those of the Java or Pekin men. There was no trace of a forehead, and the mouth came out in a regular snout. At first sight they might only be skulls of apes a little more man-like than the chimpanzee.   At least one distinguished anatomist takes this view.

But Broom, and Legros Clark, who has examined them on the spot, say that they are much more human than those of any ape, for the following reasons among others.   There is no doubt that the tailless apes, the gorillas, chimpanzees, orangs and gibbons, are much nearer to us than any of the tailed monkeys, both in structure and in mental capacity.   But they all differ from us in several ways.   They have prominent canine teeth, which the males especially use in fighting.   The hole through which the spinal cord enters the brain is set further back in the skull than in men, so that the head must be bent down to look forwards if they stand on their hind legs.   And their legs are far less specialized than our own for standing.   The heel is one of the human organs which differs most from that of the apes; and the pelvis, the bony basin which supports our abdominal organs, is also very different in men and apes.   Finally the apes have arms which may be even longer than their legs, and are used in swinging below branches instead of walking above them on all fours as the tailed monkeys do.

In all these respects the South African fossils seem to be human rather than ape-like.   Their dog teeth project little more than ours.   This means that they probably fought with their arms, perhaps holding stones or sticks.   The foramen magnum, the large hole in the base of the skull, is well forward as in man, not backward as in the apes.   No complete arm bones have been found yet, but the fragments found suggest that the arms were short.   Further the other animals whose bones are found with them in the caves were typical dwellers on the veldt, quite different from forest animals.   This means that they lived in an environment where even though they climbed trees, they would not have been able to go for miles at a time by swinging from one branch to another.   So very long arms would have been of no use to them.

Finally Broom has found not only a pelvis, but an astragalus, the bone which forms an arch at the instep between the heel and the rest of the foot, and supports the weight of the body.   In each case the form was human, and it is reasonably sure that these animals walked on their hind legs. But yet their skulls were decidedly ape-like, apart from the modification at the back due to their owners standing up. And they left no tools or anything else suggesting that they could be called men. Their stature was probably rather less than that of any living human race.

It is possible that we are descended from these animals, but rather unlikely. But they show two things. In the first place there were animals with many human characters, but not human brains. If in another hundred years no such skeletons have been found anywhere but in South Africa, we may have to admit them as probable ancestors. However, it is quite possible that similar fossils will be found in other fairly dry areas, for example in Central Asia, when people start looking for them seriously.

Secondly they show that it was possible for such animals to live, walking or running as men do, without human brains. But they were in a position to use their hands for something more skilled than holding on to branches, and if they did so, there was a selective advantage in developing their brains to control their hands. It is not at all sure that further brain development would be any advantage to an animal like a chimpanzee which already uses its hands very efficiently for grasping branches. It is certainly advantageous for an animal which is beginning to use tools, even of the crudest kind.

Moreover such an animal would gain much more from social behaviour than a chimpanzee. A chimpanzee can get away from a lion by climbing, and can move through the trees far faster than a leopard. A band of twenty chimpanzees would be no safer than a band of four. But on the ground a band of twenty, with sticks and stones, might put a lion to flight, while four could not. Combination for hunting is also much more effective on the ground than in the trees. Whether or not the species discovered by Broom are actually ancestral to man, they give us a good working idea of what the animal species, which took to using tools and became men, was like.

Broom, by the way, is almost as worthy of study as the fossils which he has collected. He is now over eighty years old, but more vigorous than many men of forty. He is a qualified doctor, and has earned his living as a doctor first in Australia and then in

South Africa, working in country districts where he could spend most of his time studying first living animals and later fossils. In South Africa he collected a series of fossil forms which show how reptiles very gradually evolved into mammals. At the age of seventy or so he took up his present line of research with complete success.

Although it was not until the age when most scientists retire from work, that he was given a university appointment, he did the work of a dozen ordinary lifetimes in his spare time. It is good to think that such a man has made the most important discoveries of our generation concerning human evolution.

3

Darwinism and its Perversions

MOST Marxists are Darwinists. Stalin was turned out of a theological seminary for reading a translation of one of Darwin's books. Nevertheless Darwinism has been used to defend highly anti-democratic ideas. The fact is, I think, that Darwin went badly wrong, not in his account of how evolution happened, but in his comments on the process.

We can state the theory of natural selection something like this, in modern terminology. If a number of animals or plants in a population carry a gene which makes them fitter than the rest of it, in the sense that on the average they leave more descendants behind them, that gene will tend to spread through the population. A gene is a structure in the cell nucleus usually, but not always, too small to see with a microscope, and handed on to the next generation by a process of copying. And of course fitness is not a mere matter of survival or fertility. An animal which looks after its young is fitter, in the Darwinian sense, than one which does not, because more of them survive to maturity. Natural selection does appear, as Darwin thought, to be the main driving force of evolution. In fact modern work has decisively confirmed its importance.

Unfortunately Darwin did not stop here. He wrote of natural selection "favouring the good and rejecting the bad", and even ventured to predict "And as natural selection acts solely by and for the good of each being, all corporeal and mental endowments will tend to progress towards perfection." Of course he realised that most lines of descent in the past had ended in extinction, but he apparently thought that this was always due to competition by "better" or more perfect species.

Marx and Engels pointed out that Darwin was severely biased by his views as a well-to-do member of the English bourgeoisie. The passages quoted show that he did not draw a clear distinction between goodness and success, or might and right.   The capitalist class begins to notice the distinction when things are going against it, as at present, though curiously enough it is still as convinced of its own righteousness as it was a hundred years ago, when only a few men like Marx saw that it was already preparing its own doom.

The fact however is that the survival of the fittest does not necessarily make a species of animals or plants fitter in any intelligible sense of the word.   It usually does so, but it may equally make it less fit.   In a great many species of mammals and birds polygamy prevails.   The strongest males have a larger number of mates, the weaker have none.   In such species one usually finds that the males are considerably larger than the females.   Thus male poultry and pheasants are larger than their hens.   In the monogamous song birds the sexes are generally of the same size.   In some highly polygamous animals such as the fur seal the disproportion is very great.   Clearly where the males fight for the females, mere size is an advantage.   It is not necessarily an advantage in the struggle with other species.   A large seal is compelled to feed on large fish, and they may not be so numerous as small ones.

The record of fossils shows that very many species have progressively increased in size, very often with a development of horns or other such weapons in the male sex.   This increase in size was often the prelude to extinction.   The large species died out, while smaller ones lived on.

Specializations of all kinds may give an advantage to their possessors.   Many of our common insects can only live on one kind of plant.   So long as the plant is common, natural selection will tend to favour those which are particularly well adapted to it.   But once a rigid specialization is established, if the plant dies out, so does the insect.

Of course very occasionally an extreme specialization opens up a new field to an animal.   For example the transformation of the front legs into wings allowed the birds and bats to conquer the air.   But usually such specialization only enables animals or plants to use a limited habitat, for example caves or cliff-faces. Sometimes these specialists have a stroke of luck.   For example, the pigeon, which is a cliff-dweller when wild, has found artificial cliffs in the buildings of London.

Even an apparently advantageous adaptation may reduce the numbers of a species. If a new gene appears in a species of insect which makes it less conspicuous to birds, natural selection is likely to make it spread throughout the species. But if the same insect is attacked by an internal parasite, so many parasites may survive that the numbers are actually reduced.

Natural selection is, I believe, the main agent of evolution. And it certainly prevents animals from losing useful organs and instincts, as they may do when domesticated. But it is a blind force, not necessarily beneficial, in the long run. I think it probable that many species have become extinct as the result of natural selection, which forced them into evolutionary paths which were blind alleys.

In the same way economic forces determine the development of human societies, as Marx first clearly saw. But whereas the earlier economists, such as Adam Smith, thought that economic competition would necessarily make all nations richer, we now see that this is not true. On the contrary, competitive capitalism, by the survival of the few "fittest" businesses, inevitably develops into monopolism. Fortunately Marxists see that there is a way out. But they have a big task to convince their neighbours, and a terribly short time for their task.

I think that future students of evolution will build on Darwin's work as Marx built on that of Smith, Ricardo and others. But this will only be done by a study of natural selection at work. Its time scale is so much slower than that of economics that we cannot hope for the necessary knowledge in one human generation. For this reason it is necessary to apply dialectical thinking to Darwinism. Until we get it, it is futile and dangerous to talk about controlling human evolution. Hitler tried to do it. Hitler is dead, but his ideas are alive, and we must be very careful to see that Darwinism is not made the basis of a new Hitlerism.

4

The Mathematics of Evolution

THE greatest difficulty in explaining science to ordinary people is that almost every part of it is becoming mathematical. The mathematics are not always very difficult. For example you do not need much more mathematics to study heredity than to study contract bridge. But you do need some.

One of the studies which is rapidly becoming mathematical is that of evolution. Darwin thought in words. His successors to-day have to think in numbers. Everyone who has gone into the evidence, which takes some years to do, believes in evolution. That is to say he believes that the animals and plants living to-day are descended from very different ones in the past, some of which have left fossils. But there is a good deal of doubt as to some of the lines of descent and an immense amount about how evolution happened.

Most biologists think an explanation based on natural selection will account for it. But some believe, with Lamarck, that acquired characters are inherited, for example that if you feed a hen well, not only will it lay more eggs, but so will its daughters. Others believe evolution is divinely guided, in spite of the fact that this puts the responsibility for the tapeworm and the tubercle bacillus on God (for there were certainly parasites long before there were any men to sin). Still others say they don't know.

The first place where mathematics come in is in fixing the time scale. This can be done by analysing radioactive minerals. For uranium and thorium gradually transform themselves into lead, which has a different atomic weight from ordinary lead, And the older a radioactive mineral the more of this special type of lead it will contain.

The next step is to measure a number of fossils carefully in order to see just how much change has occurred in, say, two million years of evolution. The results are astonishing. The teeth of horses have been getting longer for some fifty million years. Their ancestors were browsers, that is to say they ate the leaves of trees, for which they only needed short teeth. But grass is a good deal grittier than tree leaves, besides containing grit from the soil, and wears the teeth down. So a short-toothed animal could only live for a year or so on grass. It would die when its teeth were worn away. But teeth have changed so slowly that if you measure corresponding teeth from a population of fossil horses and from their descendants two million years later, although the average values have changed, there is often still some overlap. That is to say the shortest teeth two million years later are no longer than the longest two million years past.

The next step is to see if you can change the characters of a population by exposing it to natural selection under controlled conditions. This has been done with populations of flies by Dubinin in the Soviet Union, by Dobzhansky in the United States, by Kalmus in England and above all by Teissier in France. The mathematical theory of these changes is fairly complicated, and a part of it was worked out by myself before any of these experiments had been done, while Wright and Fisher have tackled some of the still more complicated problems which arise in natural evolution.

Curiously enough we know more about natural selection in man than in any other animal or plant. The reason is a simple one. One can study human beings with various inherited abnormalities and see how long on an average they live and how many children they have. One cannot do this with wild animals. White mice in captivity are just as fit as coloured ones. They live about as long and have as many children. And whites do not disappear from a mixed population. But they are less fit in the wild state, probably because they do not see as well as normal mice, and are more conspicuous to their enemies. However, one cannot study a thousand wild white mice and a thousand coloured ones, and see just how the white ones are less fit. One can make such studies on hundreds of human dwarfs or haemophilics, that is to say boys whose blood clots very slowly.

So the most immediate application of the mathematical theory of natural selection has been to human society. Unfortunately most of the ladies and gentlemen who wish to improve the human race seem to find the theory a bit too stiff. I do not blame them for finding it stiff. I do blame them for putting forward eugenical schemes without the necessary mathematics. This is as futile as trying to design a high-speed aeroplane without mathematics, and a lot more dangerous. For a badly designed aeroplane will probably only kill a few pilots and passengers. But false ideas about racial biology may lead to the death of millions, as Hitler demonstrated.

In fact the theory shows that some "racial hygiene" is possible, but that it is far less efficient than has been thought. We could prevent about half the haemophilics from being born, and about a quarter of the dwarfs. In either case we should have to interfere with human liberty to some extent. I doubt if it would be worth while. It would be still harder to stop mental defectives from being born, for the good reason that most of their parents are normal. This does not mean that it will always be impossible either to prevent the birth of such children, or to treat them so that they grow up into rational people.

To come back to evolution, I think it has been proved that natural selection is an effective agent, and will explain a very great deal of what has happened. But some changes are certainly harder to explain than others; and I think it is still an open question whether all evolutionary change can be explained in this way. What I am sure of is that it is as useless to argue about some of these doubtful cases in words as to argue in words about whether or not an aeroplane will ever fly faster than sound.

The fact that science is getting more mathematical is one of my main difficulties in explaining it. The remedy is for children to learn more mathematics, which they could do if mathematics were brought into relation with real life, instead of with ridiculous problems about the price of eggs, which is controlled anyway. But till my readers know more mathematics, I have to write more dogmatically than I like.

X

GREAT MEN

Langevin

WHEN Paul Langevin died in 1946, I was asked to write an obituary. I had to refuse, for the good reason that I did not know enough about his scientific work. I knew that he had advanced many branches of physics and that he was one of the two foreigners to whom the Royal Society had awarded two of its medals. I was far from clear as to exactly what he had done. Only by reading the last number of *La Pensée,* the great French review which he founded, have I been able to find out the measure of his achievements.

In 1897 he came over to Cambridge with a scholarship from the City of Paris to work under J. J. Thomson, who was carrying out the research on electric conduction through gases which led to the discovery of electrons, and his first ten papers were records and interpretations of experimental work on this subject.

In 1905 he published three theoretical papers on magnetism, on relativity, and on the movements of molecules in gases. I think other physicists admired Langevin's work on magnetism above anything else which he had done. Magnetism is an example of what is called a co-operative phenomenon. The individual atoms in a magnetized iron bar are no different from those in an unmagnetized one. Nor is their arrangement different. But they have a tendency to face in the same direction, and as each one is a little magnet, the bar becomes a magnet too.

So much had been guessed for a long time. Langevin was the first to approach the problem dialectically, though as far as I know he was not then a Marxist. He looked for a conflict between the influences making for a regular arrangement of atomic directions and something else. He found the antagonist in heat. The hotter the iron bar the more the atoms will be jostled out of their agreement in direction, and the less will be the amount of magnetism produced by a given current. This, and a number of other similar facts, were known before Langevin was able to calculate what happened in a number of special cases, and a large amount of experimental work by others verified his calculations.

He also predicted a quite new phenomenon, namely that when a body is magnetized, its temperature rises, though often only by a few hundredths of a degree. This was found to be the case, and is now the basis of the method used for producing the greatest extreme of cold so far reached. Gases are liquefied by making them do work in expanding rapidly. The liquefied gases are cooled still further and even frozen by letting them evaporate. A temperature is reached when they can evaporate no more. Certain crystals are then put in a magnetic field and cooled down as far as possible. The magnetic field is then taken away, and they cool down still further. A large fraction of the small amount of heat left in them is used up in destroying the regularity produced by the magnetic field. Thus the study of the conflict between temperature and magnetism gave not only a more exact theory of magnetism, applicable in principle to other properties of matter, but a new technical method which may be the basis of industries a generation hence.

The theoretical work on relativity was the first of a series which not merely confirmed Einstein's work but extended it considerably. Einstein, in his obituary, wrote of Langevin: "It seems certain to me that he would have developed the special theory of relativity had this not been done elsewhere; for he had clearly recognized its essential points."

What he did was to apply it to chemistry. It had long been known that the atomic weights of the elements were not exactly whole numbers when that of hydrogen is taken as a unit. In 1913, Langevin suggested that this was due to the fact that energy has weight and mass. This principle could not be applied correctly till Aston had weighed atoms correctly to one part in a thousand, ten years later. It is now accepted by all physicists. If we take the atomic weight of hydrogen as one, those of the two different kinds of iron atom are not exactly 54 and 56, but nearly one per cent. below these values. This is because if we could build up iron atoms from hydrogen a lot of energy would be lost, and this energy has weight.   A similar calculation from atomic weights gives the energy liberated by an atomic bomb.

Now comes a rather amazing coincidence.   Every communist ought to read Jack London's novel *The Iron Heel* in which he predicted the coming of fascism.   He was quite right regarding its successful splitting of the workers' movement, and its extreme cruelty.   Not being a Marxist, he did not see that it would be unstable, and could not last even for a generation, instead of the centuries which he predicted. The book is supposed to be written by the widow of a socialist leader, Everhard, who had been executed by the fascists.   Her father, a Californian professor, and himself a convert to socialism, had discovered the identity of matter and energy.

Langevin cannot be said to have done this.   He only showed that some of the weight and mass of atoms was that of the energy in them.   It is not yet sure that all the weight and mass can be converted into any form of energy.   However, his daughter Hélène married Jacques Solomon, a physicist, and one of the many communists who were shot by the Gestapo during the German occupation.   Langevin was imprisoned in 1940, but later released, after which he escaped to Switzerland when again threatened with arrest.

From 1905 onwards his publications became more and more mathematical, and it looked as if he had abandoned experimental work. However, during the First World War he started research on the production of beams of supersonic vibrations in water. Ordinary sound waves spread rapidly, and turn corners easily. But trains of waves too shrill to hear behave much more like light, and can be used like a searchlight beam or a headlamp to detect obstacles under water, or even submarines.   Both as a source of such waves and for detecting them he used the phenomenon of piezo-electricity discovered by the Curie brothers. A quartz disc in an alternating electric field contracts and expands as the field changes, and will similarly translate changes of pressure into electric surges. His work in this field has not only been used under water, but in radio engineering, in the design of very accurate clocks, and in many other fields of practical work. I have no space to mention his work on chemistry, on radioactivity, on units of measurement, and even on why the sky is blue.

He took teaching very seriously. Although he wrote a number of books, he never published his courses of lectures on physics; however there can be no doubt that they had a very great influence in bringing the teaching of this subject up to date, and above all in welding it into a unity rather than a series of branches such as light, heat, sound, electricity and magnetism.

As early as 1925, he was one of the founders of a movement which led to the foundation in 1932 of the *Université ouvrière,* that is to say the Workers' University, of Paris. Daladier suppressed it in 1939, but it was born again in 1945 as the *Université nouvelle.*

He was also an ardent supporter of the *Ligue des Droits de l'Homme,* an organization which did similar work in France to Civil Liberties in Britain. Up to 1936, he showed some leanings towards an extreme pacifism. If so, Franco cured him, for he was one of the most ardent supporters of the Spanish Republic in its glorious but unsuccessful struggle. It was natural that such a man should become, as he did, a member of the Communist Party.

His scientific work was remarkable as demonstrating the unity of practice and theory. Certainly no British physicist since Thomson, perhaps none since Newton, has combined such fundamental advances both on the technical and theoretical sides of his science. When his full biography can be written without hurting the feelings of others yet alive, it will appear that his emotional life, without any detailed reference to his scientific or political work, might have furnished, and may yet furnish, the material for one of the world's great books. He was, in fact, an all-round man. We need such men to-day.

Hopkins

SIR FREDERICK GOWLAND HOPKINS, who died in May 1947, was the founder of modern biochemistry. But his work was relatively little known outside scientific circles, perhaps because few of his discoveries were of the kind that made headlines at the time, though they opened up new fields of research. What is more surprising, he only published about six papers of first-rate importance. Nevertheless, every biochemist to-day takes for granted a point of view which was revolutionary heresy when Hopkins first invented it; and this point of view has been the major influence in biochemistry in the last 30 years.

Hopkins was a great analyst. He worked for some years in the laboratory of the Government analyst, and played a part, of which he seldom, if ever, talked, in securing the condemnation of a number of poisoners. The discovery which made his reputation was based on an analytical method. It had been long known that most proteins gave a colour reaction with acetic acid. One day Hopkins was teaching a class of students how to get this and other reactions. One of the students, S. W. Cole, failed to get it. Some teachers would have told him to try again. Hopkins checked his work, and got no colour. He settled down to find out what was wrong, and discovered that the reaction was not due to acetic acid at all, but to glyoxylic acid, which is a common impurity in laboratory acetic acid. When they used glyoxylic acid the reaction became much stronger, and he and Cole set out to discover what was the component in proteins which gave it. They finally isolated a substance called tryptophane which is present in some but not all proteins, and which he later showed to be a necessary constituent of any complete diet.

To prove this, he had to design a diet for rats on which they lost weight unless a tiny fraction of tryptophane was added, but grew normally when the addition was made. It was not sufficient to give the rats proteins, fats and carbohydrates, as the books of fifty years ago stated. They needed something else, which Hopkins called accessory food factors, and obtained from milk. A worker who thought he had purified one of these factors gave it the name of vitamin, and this word caught on, in place of Hopkins' more accurate expression. But the two articles he wrote, in which he proved that rats need tryptophane and a group of unidentified substances found in milk, have been models for all later work. It is noteworthy that his rats did not get particularly ill. But those kept on the complete diet grew so well that a loss of weight by the others was sufficient evidence that their diet lacked an essential constituent.

Hopkins never isolated any of the vitamins, though St. György prepared one of them, ascorbic acid, in his laboratory; but he had, quite unintentionally, gone a long way to work out the structure of one of them. For he collected butterflies and matchboxes. From the wings of some butterflies he isolated a peculiar pigment called pterin, whose structure he partially worked out. Forty years later it was found to be a constituent of folic acid, one of the vitamins, and also of the substance which cures pernicious anaemia in men.

His other great discovery was his proof, with Fletcher, that lactic acid is formed in muscles when they contract. Since then scores of other such substances have been isolated, and indeed the chemistry of muscular contraction is fairly well understood. But this discovery was the first vindication of Hopkins' guiding principle, that it was possible to trace the whole set of transformations which a chemical substance underwent while passing through a living animal or plant.

Fifty years ago physiologists thought that food reaching living cells was somehow incorporated into living protoplasm, and that it was useless to apply ordinary chemical ideas to intermediary metabolism, which is the name given to the pattern of chemical changes in the living cell. Hopkins believed that chemical principles could be so applied. His work on diet was guided by the idea that an animal needs certain compounds which it cannot make itself, and that a complete diet contains enough of each of them. Our rationing system is based on this simple idea.

In 1922 he became the first professor of biochemistry at Cambridge, and chose me as his second-in-command, perhaps because my work had been so very different from his own that he hoped that between us we should cover a pretty wide field. I think I disappointed him by deserting biochemistry, but I did at least learn some of his ways of thinking and apply them to genetics.

As a chief he was too kind and too modest. He would not plan other people's work or get rid of people who were wasting their time and his. On the contrary, he could be relied on to help his weaker pupils in their personal difficulties. I have never known a chief who was more universally loved by his subordinates. Fortunately some of them worked on the lines which he had laid down. In particular, Quastel and Stephenson found out a very great deal about the chemical processes going on in living bacteria, and laid many of the foundations for recent work on chemo-therapy.

He never produced a theory of the life process. The laboratory published a humorous annual, *Brighter Biochemistry*, to which he contributed regularly. One of his articles, on biochemistry a hundred years hence, perhaps revealed what he really thought. The biochemists of the twenty-first century were applying higher mathematics to "psychoids" in the liver and other organs, while the physicists were mainly engaged in extremely accurate measurements of new properties of matter revealed by the biochemists. The framers of the current Soviet five-year research plan have a somewhat similar idea.

Rather late in his life his contemporaries recognised Hopkins' greatness, and within a few years of his becoming a professor he received the presidency of the Royal Society, the Nobel prize, the Order of Merit, and other distinctions. He never took these honours quite seriously, and was at his best explaining to juniors in an after-dinner speech that the expectation of life of a Copley medallist of the Royal Society was only three years, so as he had received the medal promotions would soon occur. He was no politician, but an enthusiastic supporter and at one time president

of the National Union (now the Association) of Scientific
Workers.

We can do nothing more for Hopkins, but we can try to
organise society so that scientific workers as socially modest and
as intellectually bold as he get their chance in life earlier than
he did, and can contribute as fully as possible to progress.

3

Lea

LAST month (June, 1947) Dr. D. E. Lea, who had just been appointed Reader in Radio-Biology at Cambridge University, died as the result of a fall from a window. His death is not only a loss to pure science. It may conceivably entail your death or mine. For Dr. Lea was one of those workers engaged in investigating the action of radiations on living things who had refused to be entangled in the net of secrecy which spreads out in every direction round the atomic bomb. From what he said about such secrecy it is fairly clear that he would have refused to do secret work. So many discoveries which he might have made would have been available for the protection of the general public.

His book *The Action of Radiation on Living Organisms* is certainly the best summary available in any language of what happens when X-rays, gamma rays, or rapidly moving electrons, alpha particles, neutrons, and so on, penetrate living tissue. The effects of all these agents are very similar, because when a gamma ray or X-ray is stopped by an atom, a high-speed electron is shot out, and it is this which causes most of the damage before coming to rest.

The damage seems to be of two rather distinct kinds. In the first place cells which have been heavily irradiated cannot divide for some time, and often die when they do so. This is why X-rays are far more deadly to men or any other vertebrates than to adult insects. An insect is, in one respect at least, much more of a machine than is a man or a mouse. It is made of parts which are not replaced, and dies when they wear out. Its skin is hard, and no more is formed after its last moult or its emergence from a pupal case. Whereas our skins are constantly being replaced. So a dose of X-rays which will not harm an insect will cause serious skin burns in men, because the skin cells cannot divide to make new skin as the old skin wears away. We are also constantly replacing our blood corpuscles, by the division of cells in the bone marrow. So anaemia is another consequence of over-radiation. On the other hand cancer cells divide very frequently, so it is often possible to kill a cancer with X-rays while sparing the normal organs round it whose cells are not dividing.

X-rays and quickly moving particles have another effect to which Lea devoted more time. When a cell divides most parts of the new cell are made afresh. But some essential parts are copies of the corresponding parts of the old cell. If one of the uncopied parts is damaged this causes no permanent changes provided the cell survives at all. But if one of the copied parts is affected it is copied in its changed condition. And if the cell in question happens to be an ancestor of germ cells the change may be inherited for many generations. Such changes are called mutations, and though they are generally harmful, this is not always so. In a recent article I described useful changes produced in this way in crop plants in Sweden and in the Soviet Union.

Lea's most important work was a very careful comparison of the effects of different kinds of rays and quickly moving particles. He showed that if we are considering a single "target" in a cell, say a gene responsible for producing colouring matter in a fly's eye or hairs on a barley head, we can measure the area of the target from the chance of a "hit" by one kind of particle, and the volume from the chance of a hit by another kind. Thus he got two quite independent measurements of the size of a gene, which agreed very well. He applied the same method to measuring the size of viruses, such as the virus of cowpox which is used for vaccination. He went on to consider more complicated changes, such as rearrangement of the structure of chromosomes. I had the honour of helping him with some rather tricky mathematics which enabled him to get slightly more accurate results in this case.

I do not know what he would have done next. He might have investigated the possibilities of protection from these effects by chemical agents. This would be very important for workers on artificial radioactivity and atomic fission who receive accidental injuries, and perhaps even in defending populations against atomic bombs. For example, tadpoles can be protected from the effects of X-rays by keeping them in very cold water after a heavy dose, so that their cells have a chance to recover before they divide. You cannot cool a man down much without killing him. But you can slow down cell division with sulphanilamide derivatives, and this might conceivably save his life.

Any experiments to test such a possibility would have to be extremely critical; and Lea was nothing if not critical— at least as critical of his own work as of other people's. In fact he made his most fundamental discoveries because in some of his earlier work different kinds of treatment had given results much more different than were expected on the basis of the theory on which he was working. He pointed out the contradiction, and in the course of explaining it he did a number of most important experiments, and showed how one kind of treatment measures the area of the target, and another its volume.

Most of the other British scientists on this subject are more or less gagged by "security" regulations. They may merely have been asked for advice on the protection of workers with radioactive substances, but in giving such advice they have learned facts which are secret. Now these questions will become more and more important. So will the question of the protection of the public from the waste products of factories or laboratories such as that at Harwell. Personally I believe that up till now the public is in no danger.

But I have not got Lea's knowledge of the scientific side of this work; and those who have a comparable knowledge (I do not think anyone has as much) are muzzled. The question will certainly come up before the public in the next few years. And no one will be able to advise them as Dr. Lea could have done. That is why his death is a serious matter for you and me.

4

Jeans

SIR JAMES JEANS was a very competent mathematician who applied his talents mainly to the study of gases. He worked on the theory of gases at ordinary temperatures and pressures, introducing various refinements of the simple theory which treats the gas molecules as if they were perfectly smooth elastic balls, and which gives a good approximation to the observed facts.

Even more important was his work on gases at very high and low temperatures and pressures. He showed that the solar system could not have originated, as Laplace thought, from condensations in a spinning disc-shaped mass of gas, each condensation attracting the gas in its neighbourhood and becoming a planet. However the nebular hypothesis has been revived by Weiszäcker, with additional postulates which at least partially meet Jeans' criticisms. His book *Problems of Cosmogony and Stellar Dynamics* was a landmark in the history of astronomy. It was however based on physical theories which are now known not to be quite exact. Nevertheless, no subsequent worker can conceivably neglect it.

For some time Jeans had told his friends that at the age of 55 he proposed to abandon pure science and devote himself mainly to popularisation. He did so with the greatest success. But future generations will remember him for his earlier work.

He died in 1946, and his last book, *The Growth of Physical Science** has just been published. This is a history of physics and of some branches of mathematics from the earliest times, and is very well worth reading. There are a few slight mistakes, particularly in the index, which the author would probably have corrected had he lived, but they are quite irrelevant to his main argument. To me, the most interesting parts of the book are

* Cambridge University Press, 12/6.

the quotations from Copernicus, Newton, and other great men. What they actually wrote was often very different from the summaries of their views which are usually given. In particular Newton did not plump for a corpuscular theory of light, as is often stated.

Unfortunately, the last chapter, which deals with modern developments, is hardly up to the standard of its predecessors. One reason for this is that Jeans tells us nothing of the history of the theory of probability, but suddenly introduces this notion in connexion with quantum mechanics. Now the theory of probability is something highly practical. It arose from a consideration of gambling and insurance, and has been applied in almost all branches of science. On page 335 Jeans equates probability with knowledge. This idealistic formulation is only sometimes true. If I say there is a probability of one in fifty-two that the top card in a well shuffled pack is the ace of spades, this is equivalent to stating that I know nothing about which card is there. But if I talk of the probability of future events we can only equate it with partial knowledge if we think that all future events are absolutely determined already. If we believe that human beings can make real choices, then the probability that I shall get drunk to-morrow is something quite different from the probability that the top card is the ace of spades.

In fact, as is so often the case, Jeans' idealistic account of probability is only a manifestation of mechanistic thinking. If you insist on treating the universe as a machine, you will have to bring in supernatural agencies to explain the facts of ordinary experience as well as those of advanced physics. His account of modern theories of the universe is far from satisfactory. Many questions, including that of the alleged expansion of the universe, are certainly far more open than a reader of his book might suppose.

For an up-to-date discussion of this question and of theories of the universe in general, I cordially recommend Paul Labérenne's *L'Origine des Mondes*.* For one thing he devotes a whole chapter to Jeans' work, which Jeans himself, with rather undue modesty, dismisses in a paragraph. But he also describes the work of

* *The Origin of Worlds,* Editions Hier et Aujourd'hui, Paris.

Tolman in America, of Fessenkoff in the Soviet Union, of Banerji and Sen in India, and of Milne in England, to mention no others, which lead to points of view decidedly different from those of Jeans.

The book is written from a Marxist angle, and is only one of a number of excellent books on science which are being written by French Marxists. My only criticism of it is as follows. The author takes such care to avoid mathematical arguments which his readers might not be able to follow, that they may not realise the great knowledge of mathematics which is needed before one can criticize an astronomical theory, let alone produce one. I constantly get letters containing astronomical theories which are either so vague that they cannot be tested at all, or alternatively which would require years of work to see whether they agreed well enough with the known facts to be worthy of further examination.

However much we may criticize such men as Jeans and Eddington, they were first-rate mathematicians, and their theories were worked out in great detail. Labérenne is a professional mathematician, and his criticisms are based on a very considerable knowledge. In fact they go deeper than a reader might think at first sight. He also sees clearly the social background of the views held by different astronomers. Jeans, in his second chapter, sees clearly enough why slavery led to a contempt for practice which sterilized Greek science. In his sixth chapter he writes of the origin of the Royal Society, quoting Boyle's description of it as "our new philosophical college which values no knowledge but as it has a tendency to use". Unfortunately he has nothing to say about the relation between science and society in our own time.

Labérenne fills this gap. We see clearly how, for example, a French catholic writer, M. de Launay in *L'Eglise et la Science,* joined with the Nazis in attacking relativity because Einstein was a Jew, oblivious of the fact that the Jesuit Lemaître had made an important contribution to it. But as Lemaître had the bad taste to agree with a Soviet mathematician, Friedmann, he stood condemned. He explains the reasons which made so many astronomers accept rather uncritically the arguments suggesting that planets were very rare, so that it was unlikely that there were intelligent beings on other stars.   The evidence of the last three years suggests that planets are rather common.

In fact we cannot study even astronomy without remembering that astronomers are human, and therefore part of society. Labérenne never forgets this fact, and that is one reason why I hope that his book may be translated into English.

5

G. H. Hardy

PROFESSOR G. H. HARDY, who died last month (November, 1947), was probably the greatest British mathematician of his generation, and one of the greatest in the world. Like many great men, he held views and did things which do not easily go together in the lives of ordinary men.

He was a very pure mathematician. Much of his work was on the theory of numbers. For example he and his colleagues tackled the problem of the number of partitions of a given number. Consider the number three. You can express it as 3, as $2 + 1$, or as $1 + 1 + 1$, that is to say split it up in three ways. Four can be written as 4, $3 + 1$, $2 + 2$, $2 + 1 + 1$, or $1 + 1 + 1 + 1$, that is to say in five ways, and five in seven ways. But how can we find an expression for the number of partitions of any number? He finally arrived at the formula, which is fairly complicated. He then tackled similar problems, such as the number of ways in which a number can be broken up into a sum of a given number of squares, cubes, and so on.

If anyone told him that such work was completely useless, he was the first to agree. He boasted that his mathematics had never helped to kill a single man, and stated that mathematics were something like cricket, worth doing for its own sake. He was an intense admirer of cricket and cricketers. He would admit that various mathematicians had been in the first class. But he put half a dozen or so of them in what he called the Hobbs class, after the great Surrey cricketer. In actual fact his boast was untrue. To take one single example, there is a function called Riemann's Zeta function, which was devised, and its properties investigated, to find an expression for the number of prime numbers less than a given number. Hardy loved it. But it has been used in the theory of pyrometry, that is to say the investigation of the temperature of furnaces. And blast furnaces play a very important part in modern war.

Even cricket has its social functions. For example in spite of the strong resentment aroused by Larwood's bowling, it has certainly cemented friendship between Britain and Australia; and the prowess of Indian and West Indian cricketers has made some Englishmen who would not otherwise have done so respect members of darker coloured races. Hardy's pure mathematics had a social function of this kind. In 1913 an unknown Indian clerk, Ramanujan, sent him a letter containing about a hundred mathematical theorems. Hardy got him over to England, and he became the first Indian fellow of Trinity College, Cambridge, and later of the Royal Society.

Unfortunately he got tuberculosis. As he lay dying of it, Hardy visited him. He asked Hardy for the number of his taxicab. Hardy replied "1729, not a particularly interesting number." "What," replied Ramanujan "don't you realise that it is the smallest number which can be expressed in two different ways as the sum of two cubes?" ($10^3 + 9^3$ or $12^3 + 1^3$). Or so the story goes. Hardy is alleged to have said that Ramanujan was on terms of personal friendship with every number less than 10,000.

In spite of this attitude to his profession, which many readers of this article will regard as futile and reactionary, Hardy was a staunch opponent of what he regarded as injustice and superstition, a socialist and a trade unionist. I remember him making a recruiting speech for the National Union of Scientific Workers, which was of course a Trade Union up to 1927, and as the Association of Scientific Workers, is one again. He argued that science and mathematics were worth doing for their own sake. But he went on to say that although our jobs were very different from a coalminer's, we were much closer to coal miners than to capitalists. At least we and the miners were both skilled workers, not exploiters of other people's work, and if there was going to be a line-up he was with the miners.

The idea of art for art's sake or mathematics for mathematics' sake is an incomplete idea. But it is very much better than the idea of art for money's sake, or mathematics for engineering's sake, no matter how the engineering is to be used. If you really believe in art for art's sake you will soon want to change things so that everyone who wants can get a chance to practise art and to enjoy it. That means working for a society where everyone has the necessary leisure and means, in fact for socialism. That was as far as G. H. Hardy got.

The next stage is reached when the artist realises that his art can become a weapon for socialism, and be all the better for it. Men like William Morris, Alan Bush and, in his early plays, Bernard Shaw, got to this stage. It is certainly harder for a mathematician to do so, because mathematics only appeal to the emotions of a few people, and can only be used directly for socialism after socialism has been won.

Though I disagree with Hardy's attitude I regard it as one-sided rather than wholly wrong. It is right that every skilled worker should take pride in his or her work, particularly when it is not done to increase someone else's profits. Hardy spent his life devising intellectual tools, which he tried out on the easiest material to hand, namely "pure" numbers. Other people have used these tools for the study of mechanical systems such as telephones, and living ones, such as brains. To take an example from my own work, I have just used part of the theory of the partitions of numbers to analyse family records to see whether, on an average, certain diseases occur more often among the later born members of a family than the earlier ones.

I happen to be one of those who find an intense aesthetic pleasure in mathematics quite apart from its applications. I quite realise that this is not enough. But I also realise that those who enjoy it most are likely to do it best. So I do not feel that Hardy's attitude was wholly wrong, and I mourn a man whom I not only liked personally, but whose writings gave me some of the emotions which others derive from classical music.

6

Einstein

EINSTEIN's seventieth birthday was on 14th March (1949). He is generally recognized as the greatest living mathematical physicist. Of course, younger men are now making greater contributions to that subject than he has done in the last ten years, but no one has yet equalled his earlier work. He is best known for his work on relativity. But if he had never written a line on that subject, he would still be regarded as a scientist of the first rank.

The quantum theory was founded by Planck, but it was Einstein who made the simplest and probably the most universally valid statement about it, namely that when matter emits or absorbs light, the energy is transformed in single units. And the size of the unit is proportional to the frequency of the light. The energy of blue light is given out in bigger packets than that of red light, and that of red light in bigger packets than that of infra-red radiation, which we cannot see, but can feel as heat. That is why when we heat a metal it gives out red light before it gives out white. At a red heat some atoms have enough energy to produce red light, hardly any have enough to produce green or blue, which must be added to the red to make white.

However, his work on relativity was even more important. Let us try to explain it. Our "common sense" view is that everything has a definite shape and size, that an event happens at the same time as a class of other events, and so on. What is more some people seem to think that any denial of this view is idealism.

Let us take a simple example to show that our common sense view won't work. I drop a parcel in a steadily moving train. To me it seems to fall in a straight line, or nearly so. To you, standing on the platform as the train goes past, it seems to move in a curve called a parabola, the descent becoming steeper and steeper as time goes on. If the earth were fixed, you would perhaps be right. But as the earth is moving too, there is little to choose between the two versions.

Does that mean that the parcel has no real track, and is only something in our minds? Not a bit, says Einstein; you can give an account of the parcel's movement which will be the same for all observers. So it is probably a considerable step nearer to reality than either my account or yours. But to give such an account we have to revise our accounts of space and time. There is an interval between any two events, and there are three sorts of intervals.

The first sort of interval can be interpreted by me as entirely one of time, that is to say I may think two events happened at the same place and different times. But if you are moving relative to me you will say they happened at different places and different times.

The second sort of interval can be interpreted as entirely one of space. That is to say I think two events happened at the same time in different places. But to you they may seem to have happened at different places and also at different times.

Common sense, rather reluctantly, recognises the first kind of relation between events. We all agree that if London is spinning round the earth's axis, two events in the same room at an hour's interval can be said to be hundreds of miles apart. But it took Einstein to see that "at the same time" was just as relative to the observer as "in the same place". There is a third kind of interval between events which all observers will agree are separated both in space and in time.

Of course if he had stopped there his work would merely have been negative. But he was able to describe a framework of space-time which was the same for all observers, though they would interpret it a little differently. This at once cleared up a lot of contradictions in physics. People had tried to measure how fast the earth was moving through space by measuring the speed of light at different times of year, and had found no differ-ence. If Einstein is right, they could not hope to find one, because space has no being of its own apart from matter.

I think most physicists are agreed that Einstein's theory works very exactly so long as the two observers are in uniform motion relative to one another, like a man on a platform and a man in a steadily moving train. But things are not so simple when the speed of one relative to the other is changing, for example when the train is accelerating or slowing down. Everyone knows that acceleration generates forces, for example an accelerating or decelerating train seems to slope even when the track is flat. Einstein said that the man in the moving train who thinks its floor is off the straight has a perfect right to his opinion, and on this basis he predicted that gravitation and acceleration would have similar effects.

In particular light should be bent by a very strong gravitational field. This prediction was verified by Eddington during an eclipse of the sun in 1919. What is more, it was bent to the extent which Einstein had predicted. More and more other predictions came off. Einstein said that a body in motion relative to a balance was heavier than the same body at rest. So is a body with potential energy. Your watch weighs more when wound up than when run down. The amount of energy in a watch is much too small to weigh by methods at present available. But the amount of energy in a large number of radioactive atoms is enough to make them weigh distinctly more than the products formed when they split up. And this energy has been weighed.

However, the general theory of relativity, that is to say the theory applied to systems whose parts are not in uniform motion relative to one another, is not complete. When one attempts to apply it to events which are very far apart in space or time it yields results which are probably incorrect. There is nothing surprising in this. One only approaches the truth by steps. Einstein made a very big step, but he is much too good a physicist to think that he has made the last one.

Of course Einstein's theories can be interpreted idealistically, and he has sometimes done so himself, though never completely. There is a measure of truth in the idealistic interpretation. The idealists say that what we call the material world only exists in our minds. A follower of Einstein would say something like this. Events, such as human births and deaths, chemical changes, or solar eclipses, are real enough. But the framework of space and time, into which we try to fit them, is partly our own construction. There is a real set of relations between events. But different people interpret it in different ways. I say the parcel fell in a straight line, you say it fell in a curve. Each of us was giving a one-sided account of a track in space-time.

Reality is more complicated than we think. But that does not mean that things aren't real. On the contrary one might say they are more real than any isolated observer could have imagined. Only by the social act of comparing the experiences of different observers can we make the important step towards truth which Einstein was the first to make.

XI

CONTROVERSIAL

Auld Hornie, F.R.S.

Out of the Silent Planet, Perelandra, That Hideous Strength.
(John Lane 1938-1945).

MR. C. S. LEWIS is a prolific writer of books which are intended to defend Christianity. Some of these are cast in the form of fiction. The most interesting group is perhaps a trilogy describing the adventures of Mr. Ransom, a Cambridge teacher of philology. In the first volume Ransom is kidnapped by a physicist called Weston and his accomplice, Devine, and taken in a "space-ship" to the planet Mars, which is inhabited by three species of fairly intelligent and highly virtuous and healthy vertebrates ruled by an angel. Weston wants to colonize the planet and Devine to use it as a source of gold. Their efforts are frustrated, and they return to earth bringing Ransom with them.

In the second volume the angel in charge of Mars takes Ransom to Venus, where he meets the Eve of a new human race, which has just been issued with souls. Weston also arrives, allows the Devil to possess him, and acts as serpent in a temptation of the new Eve. Ransom's arguments against the Devil are inadequate; so he finally kills Weston, and is returned to earth by angels, with thanks for services rendered.

In the final book two still more sinister scientists, Frost and Wither, who have given their souls to the Devil, are running the National Institute of Co-ordinated Experiments. Devine, now a peer, is helping them. The only experiment described is the perfusion of a severed human head, through which the Devil issues his commands. They are also hoping to resurrect Merlin who has been asleep for fifteen centuries in their neighbourhood. Their aim appears to be the acquisition of superhuman power and of immortality, though how this is to be done is far from clear, just as it is far from clear why a severed head perfused with blood should live longer than a normal one, or be a more suitable instrument for the Devil. However Mr. Ransom is too much for them. He obtains the assistance not only of Merlin, but of the angels who guide the planets on their paths, and regulate the lives of their inhabitants. These angels arrive at his house, whose other inhabitants become in turn mercurial, venereal (but decorously so), martial, saturnine, and jovial, but fortunately not lunatic. Merlin and the angels smash up the National Institute and a small university town, Frost and Wither are damned, and Ransom ascends into heaven, bound for Venus, where he is to meet Kings Arthur, Melchizedek, and other select humans who escape death. One Grammarian's Funeral less, in fact.

The tale is told with very great skill, and the descriptions of celestial landscapes and of human and non-human behaviour are often brilliant. I cannot pay Mr. Lewis a higher compliment than to compare him with Dante and Milton; but to make the balance fair I must also compare him with Rolfe (alias Baron Corvo) and Velenovsky. Dante and Milton knew the science of their time, and Dante was well ahead of most of his contemporaries in holding that the earth was round, and that gravity changed direction at its centre; though Milton hedged as to the Copernican system. Mr. Lewis is often incorrect, as in his account of the gravitational field in the space ship, the atmosphere on Mars, the appearance of other planets from it, and so on. His accounts of supernatural intervention would have been more impressive had he known more of nature as it actually exists. Of course the reason is clear enough. Christian mythology incorporated the cosmological theories current eighteen centuries ago. Dante found it a slight strain to combine this mythology with the facts known in his own day. Milton found it harder. Mr. Lewis finds it impossible. Mr. Lewis is a teacher of English literature. The philologist Ransom reminds me irresistibly of the idealized Rolfe who becomes pope as Hadrian VII, though of course it is even more distinguished to escape death by ascending into heaven than to become a pope. Velenovsky (whose name is not so well known) was a Czech botanist who discovered a new species of primrose in the Balkans, and called it *Primula deorum,* the

primrose of the gods.   With such a name one might expect a
plant even nobler than the purple giants of the Himalayas and
Yunnan.   Unfortunately it is a wretched little flower, which
will not bear comparison with any of our four British species.
In his attempts to defend Christianity, Mr. Lewis has also de-
fended the beliefs in astrology, black magic, Atlantis, and even
polytheism, for the planetary angels are called gods, perhaps in
deference to Milton.   Many sincere Christians will think that he
has done no more service to Jesus than Velenovsky to Jupiter.

As a scientist I am particularly interested in his attitude to my
profession.   There is one decent scientist in the three books, a
physicist who is murdered by the devil-worshippers before we
have got to know him.   The others have an ideology which
ranges from a Kiplingesque contempt for "natives" to pure
"national socialism", with the Devil substituted for the God
whose purposes Hitler claimed to be carrying out.   As a matter
of fact very few scientists of any note outside Germany and Italy
have become fascists.   In France only one, the engineer Claude,
did so, though the Catholic biologist Carrel came back from the
U.S.A. to support the Vichy government.   A very much larger
fraction of the clerical, legal, and literary professions bowed
the knee to Baal.

Weston is recognizable as a scientist; Frost and Wither, the
devil-worshippers, are not.   They talk like some of the less
efficient of the Public Relations Officers who defend Big Business,
and even Mr. Lewis did not dare to assign them to any particular
branch of science.   At a guess I should put them as psychologists
who had early deserted the scientific aspect of psychology for its
mythological developments.

Mr. Lewis's idea is clear enough.   The application of science
to human affairs can only lead to hell.   This world is largely run
by the Devil.   "The shadow of one dark wing is over all Tellus,"
and the best we can do is to work out our own salvation in fear
and trembling.   Revealed religion tells us how to do this.   Any
human attempts at a planned world are merely playing into the
hands of the Devil.   Auld Hornie, by the way, to use the pet name
which the Scots have given him, perhaps in thanks for his attacks
on the Sabbath, has been in charge of our planet since before

life originated on it.   He even had a swipe at Mars, and removed
much of its atmosphere.   Some time in the future Jesus and the
good angels will take our planet over from him.   Meanwhile
the Church is a resistance movement, but liberation must await
a celestial D-day.   The destruction of Messrs. Frost and Wither
was only a commando operation comparable with the bombard-
ment of Sodom and Gomorrah.

In so far as Mr. Lewis succeeds in spreading his views, the
results are fairly predictable.   He will not have much influence
on scientists, if only because he does not know enough science
for this purpose.   But he will influence public opinion and that
of politicians, particularly in Britain.   I do not know if he is a
best seller in America.   He will in no way discourage the more
inhuman developments of science, such as the manufacture of
atomic bombs.   But he will make things more difficult for those
who are trying to apply science to human betterment, for example
to get some kind of world organization of food supplies into
being, or to arrive at physiological standards for housing.   In
such cases we scientists are always told that we are treating
human beings as animals.   Of course we are.   My technical
assistant keeps a lot of mosquitoes in my laboratory.   Their
infantile mortality is considerably below that of my own species
in most countries; and I hope to get it down below the level of
English babies.   But meanwhile I should be very happy if all
human babies had as good a chance of growing up as my mosquito larvae.   Mr. Lewis is presumably more concerned with
their baptism, which is alleged to have a large influence on their
prospects after death.

More and more, among people who think about such matters,
the division is appearing between those who think it is worth
while working for a better future (which, since the various
members of our species now form, for some purposes, a single
community, must be a better future for all mankind) and those
who think that the best we can do is to look after our immediate
neighbours and our noble selves.   Clearly anyone who believes
that he or she stands to lose by social changes will be pleased to
find arguments to prove that they are impracticable or even
devilish.   So Mr. Lewis is a most useful prop to the existing

social order, the more so as his Martian creatures seem to practise some kind of primitive communism under angelic guidance; so a good Lewisite can get a full measure of self-satisfaction from condemning capitalism as a bye-product of the fall of man, while taking no concrete steps to replace it by a better system.

It is interesting to see how Mr. Lewis' ideology has affected his writing. He must obviously be compared with Wells and Stapledon, rather than with the American school of "scientifiction" which is a somewhat lower form of literature than the detective story. The criteria for fictional writing on scientific subjects are similar to those for historical romance. The historical novelist may add to established history. He must not deny it. He may describe the unknown private life of Hal o' the Wynd, or Fair Rosamund. He must not contradict what little is known about them without sound reason given. In a scientific romance new processes or substances may be postulated, for example Cavorite which is opaque to gravitation, or animals which reproduce by clouds of pollen. But apart from special cases our existing knowledge of the properties of matter should be respected. Wells occasionally broke this rule; for example, the giants in *The Food of the Gods* would have broken their legs at every step; but much may be forgiven a pioneer. Stapledon is more scrupulous. Lewis' contempt for science is constantly letting him down. I wish he would learn more, if only because if he did so he would come to respect it. I do not complain of his angels or "eldils". If there are finite superhuman beings they may well be as he describes. I do complain when, in the preface to *The Great Divorce,* he writes "A wrong sum can be put right: but only by going back till you find the error and working afresh from that point, never by simply going on". I happen to be an addict of iteration. For example I have recently had to solve the cubic equation:—

$7009x^3 - 7470x^2 - 7801x + 516 = 0$. This equation arises in the theory of mosquito breeding.

Writing it as $x = \dfrac{516 - x^2}{7801}\left[1 - \dfrac{x\,(331 - 792x)}{7801} \right]$

I put $x = \cdot 06$ on the right-hand side, and get $x = \cdot 0629$ as a better approximation. Then I substitute this value on the right-hand side, and so on, finally getting $x = \cdot 06261$. If I make a small mistake it gets corrected automatically, and may even speed up the approach to the final result. I think the process of solving a moral problem, for example of arriving at mutually satisfactory relations with a colleague, is a good deal more like iteration than the ordinary method of solving such equations.

If Mr. Lewis would learn mathematics and science he might change his views on other matters, for he is intelligent enough to make some very awkward if unconscious admissions. For example the sinless creatures on Mars had a theology but no religion. They believed in a creator and an after-life, like Benjamin Franklin and other great rationalists, but during a stay of several months among them Mr. Ransom reported no religious ceremonies, or even private prayers. Their conversations with passing angels, or "eldils" whom they occasionally saw and heard, were no more like religious acts than is turning on the radio to listen to Mr. Attlee. This is entirely what one should expect if Mr. Lewis' other premises were true. A person fully adapted to his environment would have no religion. As Marx* put it. "This state, this society, produce religion—an inverted consciousness of the world—because it is an inverted world. . . . It is the fantastic realisation of man, because man possesses no true realisation."

Again it is striking that Communism is only once mentioned in the books under review, and though in *The Great Divorce,* the narrator finds one communist in hell, he had left the party and become a conscientious objector in 1941, so perhaps the punishment was deserved, if unduly severe. I take it that Mr. Lewis, who is at least aware of the important difference between right and wrong, though he draws what seems to me to be an incorrect line between them, recognizes that communists also take right and wrong seriously and is therefore loath to condemn them radically. In consequence the conflict described in *That Hideous Strength,* which is supposed to be important for the future of humanity, lacks reality. And in so far as Mr. Lewis persuades anyone that devil-worship is any more important than other rare perversions, he is merely pandering to moral escapism by

* *On Hegel's Philosophy of Law,* 1884.

diverting his readers from the great moral problems of our day.

I fear that Mr. Lewis is too "bent", to use his own word, to become a communist. Look at his taste in grammar. In the celestial language, of which he gives us some samples, the plurals of the words eldil, pfifltrigg, oyarsa, and hnakra, are eldila, pfifl-triggi, oyeresu, and hneraki. If that is his ideal of grammar, no wonder his ideals of society are peculiar. Parenthetically I should have thought the most striking character of a language used by sinless beings who loved their neighbours as themselves would have been the absence of any equivalent of the word "my", and very probably of the word "I", and of other personal pronouns and inflexions.

Nevertheless, if Mr. Lewis investigates the facts honestly, he will probably discover two things. One is that if Christianity (in the sense of an attempt to follow the precepts attributed to Jesus) has a future, that future, as things are to-day, is far more likely to be realized within the orthodox church than the western churches. In fact Marxism may prove to have given Christianity a new lease of life. The second is that scientists are less likely than any other group to sell their souls to the Devil. A few of us sell our souls to capitalists and politicians, and Mr. Lewis may have met some such vendors at Oxford. But on the whole we possess moral and intellectual standards, and live up to them as often as other people.

I think we even do so a little more often, because we possess objective standards which others do not. One can find out whether samarium is heavier than lead, whether dogs are more variable in weight than cats, or whether trilobites or dinosaurs lived earliest. There is no way of finding out whether Crashaw was a better poet than Vaughan, or whether Shakespeare wrote the parts for his heroines to suit the leading boy actors of the moment. We also have to risk our lives in the course of our profession rather more often than writers. "The real importance of scientific war" says Mr. Frost "is that scientists have to be reserved." It is worth remembering that some of us were reserved to unscrew magnetic mines and to test a variety of rather unpleasant chemical substances on our own persons.

But my real quarrel with Mr. Lewis is not for his attack on

my profession, but for his attack on my species. I believe that, without any supernatural promptings, men can be extremely good or extremely bad. He must explain human evil by the Devil, and human virtue by God. For him human freedom is a mere choice between alternatives presented to our souls by supernatural beings. For me it is something creative, in the sense that each generation makes newer and greater possibilities of good and evil. I do not think that Shaw is a greater dramatist than Shakespeare; but some of his characters, for example Saint Joan, Lavinia, or even Dudgeon, are morally better than any of Shakespeare's characters. Good has grown in three hundred years. So has evil. I do not think that any of the popes whom Dante saw in hell had done an action as evil as that of Pius XI when he blessed fascism in the encyclical *Quadragesimo Anno*.

Mr. Lewis' characters are confronted with moral choices like slugs in an experimental cage who get a cabbage if they turn right and an electric shock if they turn left. This is no doubt one step nearer to the truth than a completely mechanistic view, but only one step. Two thousand years ago some people had got further. I find Horace's "justum et tenacem propositi virum", who is not deflected by mobs, tyrants, or the great hand of thundering Jove, a vastly more admirable figure than Mr. Lewis' saints who are

"Servile to all the skyey influences;"

though of course Cato's idea of justice was as narrow as ours will, I hope, seem two thousand years hence. But it was men with this Horatian ideal of dignity who made Rome, and men with not very dissimilar ideals who made China, which did not fall as Rome fell. Both the Roman and Chinese ideals were aristocratic, They had to be so in societies where most men and women spent much of their time as mere sources of mechanical power. To-day a society is technically possible where every man and woman can have the leisure and culture needed to take a part in managing it. Democracy is in fact a possibility, but so far it has only worked rather spasmodically. Some of us want to make it a reality. Mr. Lewis regards it as impossible. "There must be rule," says an aged and learned Martian "yet how can creatures rule themselves? Beasts must be ruled by men, men by angels and angels by the creator." (I translate several celestial words). As angels do not give most of us very explicit orders, it would seem that we should entrust our destinies to someone like Dr. Frank Buchman or the Pope, who claims to be divinely guided. If Mr. Lewis does not mean us to draw such a conclusion, what does he mean by this passage?

In practice these self-styled mouthpieces of higher powers will presumably transmit orders very similar to Mr. Lewis' broadcast talks on *Christian Behaviour*. They will probably for example condemn sodomy absolutely, but they will hedge regarding usury if they even mention it. Mr. Lewis admits that Christian, Jewish, and Pagan moralists condemned it, but points out that our society is based on it, and adds "Now it may not follow that we are absolutely wrong." If it had followed that usury was wrong, Mr. Lewis' series of radio talks might have been brought to a sudden end like one of Mr. Priestley's. I mention sodomy and usury together because Dante, who expressed the ideals of mediaeval Christianity, exposed sodomites and usurers to the same rain of flames as hell, with the difference that the sodomites could dodge them, but the usurers (or as we should say, financiers) could not. If sodomy were an important part of our social system, as it was of some past systems, Mr. Lewis would presumably wonder whether it was absolutely wrong.

The men and women who believe most in human dignity are fighting usury and every other institution which makes man the slave of money. Those who share Mr. Lewis' view are compromising with these evils in one way or another, even if they do not always attack democracy as openly as does Mr. Lewis. Any Marxist can see why this must be so, and Christian readers of Mr. Lewis might well remember St. James' statement "Whosoever therefore will be a friend of the world is the enemy of God." His books certainly have very large sales, and may have a very large influence. It is only for this reason that they are worth attacking. They can of course be attacked on many other grounds than those which I have given. But I would state my case briefly as follows. I agree with Mr. Lewis that man is in a sense a fallen creature. "The Origin of the Family" seems to me to provide better evidence for this belief than the book of

R

Genesis.   But I disagree with him in that I also believe that man can rise again by his own efforts.   Those who hold the contrary view inevitably regard the reform of society as a dangerous dream, and natural science as unworthy of serious study.   And they consequently end up by making friends with the Mammon of unrighteousness.   But this friendship, so far from qualifying them for an eternal habitation, may not even secure them a competence in this present world.   For Mammon has been cleared off a sixth of our planet's surface, and his realm is contracting in Europe to-day.   It was men, not angels, who cast him out.

More Anti-Lewisite

LEWISITE is a poisonous liquid with a poisonous vapour, called after an American chemist, Lewis. British Anti-Lewisite, or B.A.L. is a compound invented by Professor Peters of Oxford, which neutralizes its poisonous effects on men and animals, and would have been used had the Germans used Lewisite against us. Fortunately, it can also be used against other arsenic compounds than Lewisite, including the familiar poison, arsenious oxide, generally though incorrectly called arsenic.

Mr. C. S. Lewis is a fellow of Magdalen College, Oxford, which has become one of our principal defenders of Christianity. His arguments seem to me to include many which definitely muddy the stream of human thought. If I can precipitate some of them, I shall help to clear this stream, thus performing in the mental sphere a task similar to that of Peters in the chemical sphere. I shall deal particularly with Mr. Lewis' Broadcast Talks.*

The first part of these talks is devoted to proofs of the existence of God. It is rather interesting to list some of the arguments which Mr. Lewis did not use. First comes the ontological argument used by St. Anselm and others, and revived by Descartes, which is roughly as follows. We can conceive of a most perfect being. But existence is a kind of perfection. Therefore the most perfect being must have existence. Mr. Lewis allows this argument to fall by its own weight, perhaps because it might be used in an inverted form to prove the non-existence of the least perfect being, namely the Devil, in whom he believes passionately.

Nor does he set much store by any of St. Thomas Aquinas' five arguments, particularly those which depend on the alleged impossibility of an infinite series of causes, or of movers. The plain fact is that St. Thomas had not the intellectual equipment

* Geoffrey Bles, London, 10/6.

to deal with infinite series, and we have this equipment to-day. They turn out to be much simpler than finite ones. Thus, if we consider the series 1/2, 1/4, 1/8, 1/16 and so on, no one can tell me the sum of its first million terms, for the good reason that its numerator and denominator each consist of 301,031 figures. But if we revise our definition of sum to cover the sum of an infinite class, we can say that the sum of all its terms is exactly unity. Mr. Lewis makes very little use of the argument from design, which, as I have pointed out, leads, if logically pursued, to the conclusion that even the animals and plants of our own planet suggest the existence of a million or more mutually hostile designers.

His main argument is from the fact that almost all human beings recognize the existence of moral obligation. At an early stage (p. 11) he deals with the argument that different societies have, or have had, different moralities. He states that they have had "only *slightly* different moralities" (his italics). Perhaps Mr. Lewis would be only *slightly* uncomfortable in a society where cannibalism was the rule, or one in which a murderer was not punished, but was compelled to adopt the children of his victim. The plain fact is that different cultures have or have had almost every morality which is compatible with the existence of society even in its crudest form. If he points out that no society has existed in which it was thought praiseworthy to murder one's parents before they reached old age, my answer is that I don't believe in miracles, and the existence of such a society would be a miracle. Societies have certainly existed in which the killing of babies and of old people were regarded as praiseworthy acts. However, let us suppose for the moment that Mr. Lewis is right, and that moral codes show a greater agreement than is necessitated by the bare existence of society, let us see how his argument continues.

He is impressed by the fact that people are aware of the existence of moral obligations, but yet do not conform to these obligations, and that people regard one moral code as better than another. "The moment", he writes, on p.17, "you say that one set of moral ideas can be better than another, you are in fact measuring both by a standard, saying that one conforms to that standard better than the other.   But the standard that measures two things is something different from either." Before we follow Mr. Lewis' next step, let us examine this argument.   If it is formally correct, it will still be true if we alter the terms in it.   Thus, if "Socrates is a man; all men are mortal; so Socrates is mortal" is a valid argument, we can substitute "Nelly" for "Socrates", "cat" for "man" and "clawed" for "mortal", and see that it still works.   Let us apply this experimental method to Mr. Lewis' argument.   Now, "tall" is a simpler idea than "good".   We do not for example ask "tall for what?" as we ask "good for what?" and it is easier to determine whether one man is taller than another than whether he is better.   Here is Mr. Lewis' argument subjected to this simple transformation.   "The moment you say that one man can be taller than another, you are in fact measuring both by a standard, saying that one conforms to that standard better than the other.   But the standard that measures two things is something different from either."

The conclusion is obviously untrue.   One can tell that one man is taller than another without any reference to a standard of measurement, and doubtless primitive men did so and do so. There are standards of measurement, but there is no absolute standard.   If people thought as loosely about length as they do about right and wrong, Britain and France would have waged a series of religious wars between the adherents of the yard and those of the metre.   But the transformation shows us something more.   Mr. Lewis writes about measuring a set of moral ideas, a notion which I find unduly materialistic.   But his notion of a standard is a standard of moral perfection to which nobody conforms all the time.   In fact it might be possible to grade different moralities, as one can grade, say, mathematical or musical performances.   But one could not do so in terms of moral perfection.   One can say that one piece of conduct or one set of moral ideas is better than another.   But one cannot say there is a best standard.   A simple example will show why this is so. I find a man bleeding by the roadside.   I certainly ought to help him in some way.   But the help that I can give depends on my knowledge and skill.   If I know nothing about first aid I can do a little, if I have taken a first aid course I can do more, if I am a surgeon a great deal more. I must always do the best I can, and it can be argued that every one has the duty to learn some first aid, so that he can stop a bleeding artery. It can hardly be argued that everyone should learn surgery. The ideal man is doubtless skilled in surgery, psychiatry and other cognate subjects, and if Mr. Lewis is correct, can even pray with enough efficiency to pull off at least an occasional miracle. But he is useless as a standard in this case. The practical standard is not the ideal man but the man who can do a little better than myself, the man who has taken the first aid course which I didn't take, or memorized the location of the nearest telephone box, which I didn't. An absolute or ideal standard of conduct is useless. And because it is useless it is immoral, in the sense that it actually leads to a less good life than the practical standard. This is one of the main reasons why, as a matter of hard fact, religion does not produce a higher level of moral conduct in its adherents than does irreligion. It sets standards which are impossible because they are self-contradictory. I cannot learn surgery, Chinese, diving, fire-fighting, infantile hygiene, wrestling, rock-climbing, weight-lifting and all the other accomplishments which might enable me to save a life. In the same way I cannot be a moral paragon in all respects. But I could always, or almost always, have done a little better than I actually did.

Mr. Lewis finds it unintelligible that we should be dissatisfied with our actual conduct unless an absolute standard of conduct exists. He can understand it if our ancestors fell from such a standard. It seems to me quite equally intelligible if our standard is, on the whole, rising. Once a conscious being can form any idea of the future he will wish it to be in some respects more satisfactory than the present. He will realise that some of the unpleasantness of the present arises from his own past actions, and will wish not to repeat such actions in future. For example, he may wake up with a headache and determine never again to drink so much whisky. This is a very elementary type of moral decision, but it is one. The passage to altruistic conduct is a more complicated matter. But one can regret past behaviour and resolve to do better without any altruism, and the possibility of doing so without any supernatural standard is the point at issue.

Our own moral behaviour is complicated by two facts. We have a cerebral structure which sometimes generates emotions more appropriate to a primitive savage than a civilized man. And we live in a society whose customs and laws are at least several generations out-of-date in relation to its productive forces, that is to say, to the jobs on which people are engaged. For both these reasons, we are frequently dissatisfied by our own conduct and that of our neighbours. I can see no reason to postulate either a god or a devil to explain this state of affairs.

Supposing there were an extra-human, or at least superhuman, standard of morality, a doctrine which I regard for the reasons explained above as dangerous and untrue, Mr. Lewis' next point would certainly not follow. "If you look at the present state of the world" he writes on page 30, "it's pretty plain that humanity has been making some big mistake. We're on the wrong road. And if that is so, we must go back. Going back is the quickest way on." Some of our religious teachers claim (and in a few cases with justice) not to be reactionaries. Mr. Lewis can make no such claim. Now, supposing I were a performing sea-lion extremely anxious to please my keeper, and aware that I could not yet balance as many balls on my nose as he wished, it would not follow that I had made any one big mistake. Much more probably I should have made a lot of little ones. I am a critic (most people think too violent a critic) of our present social system. But I don't think it is one big mistake. I don't think it is a mistake that I should be allowed to own a toothbrush, or even a dwelling house. I think it a mistake that I should be allowed to own ten acres in the City of Westminster, though this was not unreasonable five hundred years ago when this area was open country. I think it a mistake that I should be paid to give lectures to a few students rather than make really good talking films for a larger number, but this method of teaching was quite reasonable even fifty years ago. And so on.

Supposing that the moral obligations which we recognize are the standard set by a superhuman personal being, it seems just as probable that such a being for some reason prefers us to improve our conduct gradually by learning from our own mistakes, rather than use more drastic methods to make us good.

The history of man in the last few thousand years can be regarded as a series of moral challenges to which men have responded by remodelling their conduct. Sometimes this remodelling involved the collapse of a political system, as with the Roman Empire, sometimes only its transformation, as with the decay of feudalism in Britain. Such challenges have been met more or less satisfactorily in the past. They might have been arranged by a superhuman being. However, I think they are mainly the result of changes in productive forces. Thus improvements in transport and food production made it possible for a hundred thousand or more people to live in one city, and this demanded a new code of right and wrong. Further improvements in transport made the city too small a political unit, and so on. We are up against a very severe moral challenge at the present time. If we think it came out of the blue from a supernatural being it seems to me that we are much less likely to meet it effectively than if we think that it came about through changes in industry and transport which have given us on the one hand the possibility of universal plenty in a world community, and on the other hand the atomic bomb and the long-range bomber. If we think our only course is to go back, we shall not meet it at all.

So much for Mr. Lewis' argument from moral obligation. He has a few others, perhaps rather better. For example, if the universe is not the work of a creative mind he argues that thought is merely a by-product of chemical reactions in the brain. "But if so," he asks (p. 38) "how can I trust my own thinking to be true? . . . Unless I believe in God, I can't believe in thought: so I can never use thought to disbelieve in God." Let us suppose the creator has made intelligent beings on two planets. On one they are endowed with free will, which they use to such effect that most of them, after unhappy lives, go to eternal torment after death. On the other, they behave well and live happily, either ceasing to exist when they die, or going on to eternal bliss. They are all, however, afflicted with a peculiar mental set-up which leads them to believe, when they think of such matters, that there is only a finite number of prime numbers; and a good deal of time is wasted in tabulating them, in the hope of finding the largest one. I think the second world is considerably easier than the first to reconcile with the hypothesis of a benevolent creator. In fact, if we are the work of an almighty hand, and yet with no exceptions (or possibly one exception) our moral conduct is imperfect, is it not at least highly probable that our reasoning powers are equally imperfect? As a matter of fact we know them to be so. For over two thousand years all educated men believed Euclid to have proved several propositions which he did not prove. I don't "believe in thought" as Mr. Lewis perhaps does, as a process bound to lead to truth. I believe in it as a process which fairly often does so. But if I believed in an almighty creator I should certainly believe that he could make me think anything he wished, and should therefore have no guarantee that my thought processes have any validity. The survivors of Hiroshima and Nagasaki may very well wish that the creator had induced Rutherford into logical errors when he started thinking about atomic nuclei. And if the creator exists, it is highly probable that he has deliberately made it impossible for us to think about other things which would be even more dangerous. Thus I should be prepared to reverse Mr. Lewis' statement and say that if I believe in God, I can't believe in thought.

Let me be perfectly frank. I can't give an account of thought which is any better than Mr. Lewis'. But then I know a great deal less about the universe than he thinks he knows. In particular I don't expect that anyone will be able to give even a moderately satisfactory account until a lot more is known about our brains. I don't think thought is a mere by-product of physical or chemical processes in these organs. But if Mr. Lewis has ever been anaesthetized, or even drunk, he must admit that, at least in this present wicked world, his capacity for thought depends on the chemical state of his brain. On the other hand the chemical state of his brain does not depend, except to a very slight extent, on what he is thinking. By putting a narcotic in his coffee I could alter this state so that he could no longer think. And I could do this equally well whether he were thinking of the college wine cellar or the attributes of God. For this reason I think our account of thought will have to wait for our account of our brains.　I think that when certain work now half finished is published, we shall know a lot more both about cerebral physiology and about how we do at least the classificatory part of thinking.

I think I have now gone over the main arguments on which Mr. Lewis relies to make listeners share his theories as to the existence and nature of God.　I have dealt with them in some detail because he was allowed a great deal of time by the B.B.C., and those who think otherwise are not allowed time in proportion to their numbers in the population.　And, as happened to me in July, 1947, if they want to say anything particularly effective, they are not allowed to do so.　But Mr. Lewis needs attacking particularly because of his attempts, which by no means all Christian apologists make, to attack morality in the name of religion.　"If the universe is not governed by absolute goodness", he writes (p. 31) "then all our efforts are in the long run hopeless." In other words, unless you share a large part of his beliefs, there is no point in trying to be good.　It may be that "in the long run" the human race will come to an end without handing on its ethical, intellectual and cultural achievements to any other rational beings.　This conclusion was inevitable if Newtonian physics were true.　The clock had been wound up by the creator, and was bound to run down.　If Newtonian physics are not true, and diverge a great deal from truth when long periods of time are considered, it may not be correct.　But even if it is correct, I think that it is possible so to act as to make people (including ourselves) happy.　If the universe as a whole is not governed either by good or evil, it is up to us to inject some goodness into it. And this is not a hopeless task. It is a difficult one. And those who say it is hopeless make it more difficult.

Curiously enough Mr. Lewis is as contemptuous of some of the arguments for theism which others have used, as he is of lay morality.　He does not think we can deduce the existence of a creator from the physical universe.　"In the same way", he writes on p. 21 "if there is anything above or behind the observed facts in the case of stones or the weather, we, by studying them from outside, could never hope to discover it."

This is rather startling from a religious apologist.　Two

centuries ago, Addison could say of the heavenly bodies that:—

"In reason's ear they all rejoice
And utter forth a glorious voice,
Forever singing as they shine,
The hand that made us is divine."

Mr. Lewis' inner ear seems to be as deaf as my own to this song. Kant based his theism both on the starry heavens and the moral law. Mr. Lewis' theology seems to stand on one leg only. And if, as I have tried to show, his arguments from the moral law are illogical, this means that it has not got a leg to stand on.

In fact in the long run Mr. Lewis may be working for rationalism. I think that his stories which bring in witchcraft, astrology, demoniacal possession and so on, will probably bring it home to a number of people that those who reject these beliefs are a good long way towards rejecting religion altogether. But in the short run Mr. Lewis is a danger to clear thinking, and one must turn aside from more constructive work to show him up.

XI

HUMAN EVOLUTION: PAST AND FUTURE

Human Evolution: Past and Future

I AM going to treat of man from one point of view, the biological. This is wholly justifiable provided that I do not leave you with the impression that this is the only point of view that matters. Only evil can come from forgetting that man must be considered from many angles. You can think of him as a producer and a consumer. This is fully justifiable provided that you do not think that the economic angle is the only angle. You can treat him as a thinker, as an individual, as a member of society, as a being capable of moral choice, as a creator and appreciator of beauty, and so on. Concentration on only one of these aspects is disastrous. I make these obvious remarks for the following reason. Hitler and his colleagues believed that the history of the human past could be interpreted, and the history of the human future created, on biological lines. Now the Nazis degraded most of the German people morally, and brought death and misery to a whole continent. Their biological ideas were grossly incorrect. But supposing they had been as accurate as any which we possess to-day or will possess a century hence, I believe that any attempt to reduce ethics and politics to biology would have involved a moral degradation.

A biologist can do two things besides discovering facts, such as the facts of human evolution and genetics. He can tell his fellows how to achieve ends which they desire already, such as the cure or prevention of a disease. He can tell them of possibilities at which they had not guessed, such as the possibility of making childbirth painless, or some of the possibilities of which I shall speak later. But he can never tell them what is worth doing. That is always an ethical, not a biological, question. In what follows, I shall say that I think certain things are worth doing, that it is better to be born with a normal mouth than a hare-lip, with a normal colour-sense rather than colour-blind, and so on. These are my opinions as a human being. If you disagree with them, I cannot, as a biologist, persuade you to change your opinion. But if you agree, then, as a biologist, I may be able to help you to work for the ends on which we are agreed.

Perhaps you think I have taken too long over these preliminaries. But there are those who say that any attempt to apply biology to human affairs is mere Hitlerism. To my mind, that is as stupid as to say that when a tailor wants your measurement he is treating you as a mere lump of matter and no more. However that may be, I have felt it necessary to safeguard my rear. Now let us go forward.

We have been discussing evolution as something which has happened in the past, and is happening now. I may add in parentheses that we are all convinced that it has occurred, though we differ a good deal as to how, and still more as to why, it has occurred. The very first point I want to make is the time scale of the process. Forty years ago we knew the sequence of events in our evolutionary history, but could only guess at their dates. It is as if we knew that Washington lived before Lincoln, but did not know whether Washington was born 200 or 2,000 years ago. Now, thanks to the study of radioactive minerals, we know our dates with an error generally under 10 per cent. when we are dealing with dates between about 30 million and 500 million years back. We know that somewhere around 350 million years ago our ancestors were fish, 270 million years ago amphibians somewhat like salamanders, 200 million years ago reptiles not very like any living forms, and 70 million years ago, mammals something like shrews.

Curiously enough we cannot date the last few million years quite so accurately, until we get to the last 20,000, when we have layers of mud laid down each year in water from melting ice. But we can say that *Sinanthropus,* the so-called Pekin man, lived about half a million years ago, and almost surely less than a million and more than 200,000 years. Further we can say that at that time there were no men of the modern types; so our ancestors must have been a good deal less human than any existing race, even though, as they used tools, they probably deserved the name of man.

CONTROLLED EVOLUTION AS AN IDEAL

These Pekin men had queer-shaped heads, broadest about the level of the ears instead of much higher up, brow ridges, no chins, and so on. In half a million years we have changed a bit. Certainly the difference between *Sinanthropus* and modern man is as great as that which separates many nearly related animal species (for example the coyote and wolf). It is doubtful whether it is as great as that between two nearly related animal genera (for example dogs and wolves on the one hand, and various kinds of fox on the other). In fact it has taken about half a million years for a change large enough for zoologists to give it a name with full certainty. Other animals, such as horses and elephants, of whose ancestors we have a far better record, have been evolving at about the same rate.

These facts suggest that, if we did not try to control the evolutionary process in any way, our descendants half a million years hence might differ from us, for better or worse, about as much as we differ from Pekin man, or a cat from a puma. Now at present I do not think we know how to control our evolution, even if we wanted to, or if biologists were granted all the powers which Hitler exercised for twelve years. But supposing a thousand years hence we know how to direct our evolution, and further that the vast majority of men accept this ideal, as the vast majority of Americans accept the idea of sanitation to-day, what then?

The answer is rather curious. An unaided man can walk 20 miles a day with a fair load, and keep it up indefinitely. I have walked over 50 with a rifle and a few extras, but I couldn't keep it up. At present one can easily fly 2,000 miles a day, but a 5,000-mile flight is more difficult. Roughly speaking, science has increased our speed of travel a hundredfold. It is reasonable to hope that we might speed up evolution a hundredfold if we knew enough. If so, we might achieve a change as large as I have indicated in 5,000 years, or 200 generations. This is a long time. The earliest date in human history is 2283 B.C., or just over 4,000 years ago. This is the date of a total eclipse of the sun which immediately preceded the capture and destruction of Ur by the Elamites. Five thousand years ago civilization had started in

S

Egypt, Iraq, and maybe a few other areas, but most men were savages.

Is it worth while even talking about a change which we do not yet know how to bring about, and which would take 5,000 years to accomplish if we did?   Yes, it is worth while talking about it, for three reasons.   In the first place we ought to discuss the right and wrong ways of using a power before we get it.   The world would be a far safer and happier place to-day if we had discussed how to use nuclear, or so-called atomic, energy for a century, or even a generation, before we got it.   If so, very likely almost all decent people would be agreed as to the rights and wrongs of this matter, which they certainly are not to-day.   About 2,500 years ago, the prophet Isaiah got the idea that one day all the nations of the earth would be at peace.   Isaiah's idea of universal peace was something like a Jewish world empire.   Ours is an association of friendly democracies.   The ideal has only just become possible of accomplishment because world-wide transport has been achieved.   But if visionaries had not been talking about it off and on since Isaiah's time, there would be no chance of achieving it now, when the alternative is, quite literally, destruction by fire from heaven.   Now Hitler's idea of the desirable biological transformation of man was very crude indeed.   Mine may be a little less crude.   Mr. Olaf W. Stapledon's ideas, or Prof. H. J. Muller's ideas, may be better than mine, and so on. But we shall only get at better ideas by putting up our own to be shot at, as I am doing here.

Secondly, we shall not get the required knowledge in a hurry. Even now we can only do a little to alter the inborn capacities of the next generation.   Let us begin to think about what sort of changes we want, and criticize one another's ideas, as democrats should.

Thirdly, if we put the ideal of controlling evolution before us, and try to accumulate the necessary knowledge, we may find out something even more important on the way.   Columbus set out to find a sea route from Europe to China.   A ship can get from Europe to China through the Panama Canal, but the discovery of America was a vastly more important result of his voyage than the opening of this route.

SLOW DEVELOPMENT AS A MAJOR EVOLUTIONARY TREND

Our first task will be to take a glance at human evolution, and to see how man differs from the other mammals, his nearest relatives, and how these differences have arisen. Man is an exceptionally brainy animal. The whale has a heavier brain, and the mouse has a brain which is a larger fraction of its body weight. But if we take a series of closely related animals, such as the cat, ocelot, puma, and lion, we find that their brain weight is roughly proportional to the square root of their body weight. If we rate animals on the ratio of brain weight to square root of body weight, the great and small cats, for example, are about equal, and man comes out well ahead of any other animal.

We use our brains for thinking, but it is a mistake to suppose that the brain is primarily a thinking organ. Thinking is mainly, if not wholly, performed with words and other symbols, as the Greeks recognized when they used the word logic—from *logos,* a word—for the study of thought processes. From the study of the effects of brain injuries we know what parts of the brain are most concerned in thought and language. These areas are usually in the left cerebral hemisphere in the neighbourhood of the area which controls the right hand. The human brain has two super-animal activities, manual skill and logical thought. Manual skill appears to be the earlier acquisition of the two, and the capacity for language and thought has grown up around it. If we bred for qualities which involved the loss of manual ability, we should be more likely to evolve back to the apes than up to the angels.

We develop far more slowly than any other mammal. Most mammals are mature at one year or less, a chimpanzee at about 7 years, a human being at 15 or more, while growth is not complete till over 20 years, and the skull sutures are often open till nearly 30, so that the brain can still grow. We are much more like baby monkeys than adult ones. In biological language we are neotenic, like the axolotl, a Mexican salamander which, unlike most salamanders, never comes out of the water, but breeds without shedding its larval gills. Since a little thyroid hormone will make it grow up, and for other reasons, we may be pretty

s*

sure that its ancestors came out of the water. An obvious advantage of this neotenic tendency has been that man has a very long period of learning. As regards behaviour he is the most plastic of all the animals. His behaviour patterns are less fixed by heredity than theirs and more dependent on his environment.

If this tendency continues, whether by natural processes or human design, we should expect our remote descendants to have an appearance which we should describe to-day as childish. We should expect their physiological, intellectual, and emotional development to be slower than our own. We should not expect them to be born with an overpowering urge to any particular kind of conduct, good or bad.

For example, some birds are monogamous, others polygamous. Monogamy is just one of a series of fairly stereotyped behaviour patterns. Man has evolved away from stereotyped behaviour patterns, and can be monogamous, polygamous, or celibate. How he will behave depends largely on the impact of society on him. Even a cat is comparatively plastic in its behaviour. A kitten which is brought up with mice, and does not see other cats kill mice in its first few months of life, will rarely kill them. But a child's behaviour is far less fixed in advance than a kitten's.

This feature, which is so highly developed in man, and which we call plasticity of behaviour when we look at it from outside, is called the freedom of the will when we look at it from inside. In any evolution which could be called progressive we are likely to develop it still further. Bernard Shaw, in *Back to Methuselah,* shows us a young lady emerging from an egg some thousands of years hence, and spontaneously talking very good English. She is like those birds which, without education, produce a fairly complex song characteristic of their species. I can imagine human beings bred for stereotyped behaviour patterns. Perhaps if the Nazis had won they would have tried to do so. But any such step would be a step backwards.

Man is not only the brainiest species of mammal. He is the most polymorphic and polytypic if we exclude domesticated species such as the dog. Let me explain these words. We say that a species is polymorphic when in the same area there are several different types breeding together, the differences being genetically determined. For example, the fox *Vulpes fulva* of eastern Canada has three colour types: the red, cross and silver foxes. A polytypic species has different types in different areas. Your deer mouse, *Peromyscus maniculatus,* has a grey form inland, and nearly white forms on the white beaches of the Gulf of Mexico.

HUMAN DIVERSITY DESIRABLE

Man is polytypic. For example, the peoples of tropical Africa have very dark skins and kinky hair. Those of Europe have light skins and wavy or curly hair. The pre-Columbian peoples of North America have intermediate coloured skins, but straighter hair than the Europeans. Man is polymorphic. And at least as regard to colour, the European, the most successful of the human races at the present time, is also the most polymorphic. If anyone thinks that I am exaggerating this polymorphism, he will perhaps tell me of another mammalian species apart from domestic animals in which the hair colour in a single geographical area ranges from black to pale yellow, the eye colour from dark brown to pale blue.

This polymorphism is not necessarily, or even probably, due to race mixture in the past. For example, there is no reason to think that there was ever a race all of whose members had red hair. And the skull shape is as variable in cemeteries of 6,000 years ago as in modern cemeteries.

Human polymorphism certainly extends to innate abilities as well as physical and chemical characters such as stature and hair colour. For example, I am tone-deaf. I cannot distinguish between quite well-known tunes. I am pretty sure that this defect is congenital. I am also a better mathematician than the average, and have little doubt that I have abnormally high congenital ability for mathematics. No doubt I had good opportunities of learning mathematics, but so I had in the case of music. We know very little about the reasons for variation in human achievement, but we know enough to be reasonably sure that inborn differences play a great part in determining very high and low levels of achievement.

I believe that this psychological polymorphism has been a major reason for the success of the human species, and that a full recognition of this polymorphism and its implications is an essential condition for its success not only in the remote future but in our own lifetime. Let me make my meaning clearer. One of my colleagues, a man of greater manual ability than myself, and very likely of equal or greater intellectual ability, is also a musical executant who could have been a professional musician. If I had his musical gifts I might devote as much time as he does to music, at the expense of my scientific output. It is quite possible that my tone-deafness is an advantage not only for society but even for myself; though such a limitation would almost certainly be undesirable if my probable span of socially useful life were 400 years instead of 40.

THE POLITICAL IMPLICATIONS OF HUMAN DIVERSITY

I will now make a definition. Liberty is the practical recognition of human polymorphism. I hasten to add, because I recognize that your brains work differently from my own, that few of you will accept this definition. That society enjoys the greatest amount of liberty in which the greatest number of human genotypes can develop their peculiar abilities. It is generally admitted that liberty demands equality of opportunity. It is not equally realized that it demands a variety of opportunities, and a tolerance of those who fail to conform to standards which may be culturally desirable but are not essential for the functioning of society. If I lived in the Soviet Union I should not find its political and economic system irksome. I should be, and have been, irked by the assumption often made there that any cultured man enjoys listening to music and playing chess. If a nation were a pure line there would be little scope for liberty. Everyone between 45 and 50 would want so many hours a week at the movies, so much (or so little) liquor per week, and so on. These would be provided by rationing, as our needed food calories are provided in England, and everyone would be equally happy. There would be no freedom, no deviants, and no progress.

We are polymorphic not only in our aesthetic but in our

intellectual abilities.   Ways of describing the world as different as analytical and projective geometry may be equally true, even if at present one human mind cannot accept more than one of them at a time.   Last year I saw for the first time Rubens' and Breughel's great picture of Paradise at The Hague.   As a geneticist I noted with interest that the guinea pig had been created with at least three genes recessive to the wild type.   But I was even more struck by the fact that the Tree of Knowledge was infested by only one serpent but no fewer than four parrots. Maybe in the long run the parrots are more dangerous than the serpent.   Certainly we must so far recognize polymorphism as to realize that our own formulations of knowledge are not unique.

Domesticated animals such as dogs are more polymorphic than man.   But greyhounds and sheepdogs differ only because they are reproductively isolated.   The Indian caste system was an attempt to divide society into a set of reproductively isolated groups each with its peculiar function.   This system broke down, as I believe and hope that any such system would break down. I believe that when our descendants plan the genetical future of man they will have to plan for high polymorphism without reproductive isolation.   I don't know how they will do it. Fortunately I shan't have to do the planning.

Man is also polytypic.   This does not mean that any two races differ as much in intellectual, aesthetic, or moral potentialities as they do in colour.   The darkest European has lighter skin than the lightest Negro.   There is no overlap.   But even in a society where Negroes have poor opportunities of education the most cultured Negro is far more cultured than the average European, let alone the least cultured one.   Nevertheless, polytypicism has so far been an advantage to humanity.   Without postulating any over-all superiority of one race to another, we can be fairly sure that some desirable genotypes are commoner in one people than in others and that this difference is to some extent reflected in its achievements.   For example, the genotype needed for long-distance runners is relatively frequent among Finns, that needed for short-distance runners among American Negroes. Doubtless the same is true for the genotypes needed for cultural achievements.

In the past a given people at a given time has usually specialized in a few fields of culture. Thus, potential mathematicians had little chance in mediaeval Europe, but potential architects had a good chance. Very likely the contributions of a people to our common culture depend considerably on the genotypes available in it. If so, it is certainly desirable that, until all peoples have reached such a stage of liberty that rare but desirable genotypes can develop their faculties everywhere, man should remain polytypic.

If, however, 10,000 years hence we combine extreme tolerance with a psychology which will enable us to pick out human abilities at an early age, then I can see no need to foster or preserve polytypicism—though it may be desirable to do so for reasons which are not obvious at the present time. In discussing polymorphism we must not forget sex dimorphism—that is to say, the innate differences between the sexes. It is curious that in our existing society most men try to diminish them by removing their beards, while women try to exaggerate them by the use of cosmetics and other devices. *Sinanthropus* and related types seem to have been much more sexually dimorphic than ourselves. So it looks as if men conformed better than women with the evolutionary trend. It is not clear whether this trend should be encouraged to go much further. It is essential that the sexes should understand each other, but a certain difference in intellectual and emotional reactions may well be sociably desirable.

The Evolution of the Meek

To sum up, I think that in the last million years man has become more cerebral, more neotenic, and more polymorphic. I think it probable that these are desirable evolutionary trends, while I suggest that judgment should be reserved concerning polytypicism and sexual dimorphism. Others will doubtless say that I have left out the one essential: namely, a bias towards their own canon of behaviour, whether moral, religious, or political. However, I have at least given reasons why I believe that any hereditary fixation of behaviour patterns is undesirable.

Even this moderate list of desirable qualities gives us food for thought. If it were shown, for example, that the median intellectual performance of English children at the age of 15 were diminishing, and that this was not due to environmental changes, this fall could be due either to the fact that, on the whole, we were reaching a lower intellectual level at maturity, which would be undesirable; or that we were reaching the same level as our ancestors, or a higher one, but reaching it more slowly, which would be desirable.

How are we to achieve these ends? I do not know. We do not know in detail for what human characters we want to breed. The experience of animal husbandmen will not help us much, for several reasons. We do know that the domestic breeds have been selected for highly specialized performances. But in gaining desired qualities they have lost others which would be desirable in a different context. The greyhound cannot hunt by smell, the dachshund is a poor runner, the husky is ill adapted for city life, and so on.

I also think that the history of man's ancestry, as revealed by the geological record, should make us a little cautious. If a Martian zoologist who knew no more than we do now about evolution had been asked to pick the most progressive vertebrate at any time in the past, I think he would very rarely have picked on the line which was destined to give rise to man. During most of the last quarter billion years they have been pretty small, inconspicuous, and unspecialized animals. Looking at the Jurassic and Cretaceous mammals, and most of the Tertiary Primates, one might be inclined to summarize the evolutionary story as "Blessed are the meek, for they shall inherit the earth," and perhaps to suggest that peoples such as the British in the nineteenth century and the American in the twentieth, who have been successful in war, are dead ends from the evolutionary point of view.

However, if we go back to the Permian, we find that our ancestors were large and progressive reptiles. No one who looks at the skeleton and particularly the teeth of such a beast as *Cynognathus,* which was not far from our own ancestral line, could possibly class it as meek. Why the descendants of large predatory theromorphs became small and vegetarian is very far from clear. It is possible that the giant forms discovered by

Dr. von Koenigswald were actually our ancestors. If so, our ancestors a million years ago were monsters who could have torn up a tiger with their bare hands. In that case there was a second occasion on which our ancestral line tried physical dominance and gave it up again. It is also possible that the giants of Java and China were side branches from the human line, and represent an unsuccessful evolutionary experiment.

NEGATIVE EUGENICS

After all this caution, I believe we can make a start. Whatever else we may want our descendants to be, we do not want them to be blind, deaf, paralyzed, or brittle-boned. Now these conditions are sometimes due to dominant genes, which can be prevented from spreading further by negative eugenics. At first sight it might be thought that these genes could be eliminated. For example, in many pedigrees of juvenile cataract, affected persons pass on the gene for cataract to about half their children, and it very rarely skips a generation. It has been said that were they all sterilized, these conditions would be abolished. This was one of the ideas behind Hitler's racial hygiene laws.

The idea is false, because these harmful genes constantly reappear as the result of mutation. Occasionally two normal parents will have a child with a harmful dominant gene, which is then handed on until natural selection or negative eugenics puts an end to its career. The two processes are roughly in equilibrium. Thus, achondroplasic dwarfs have about one fifth the fitness of normal people. That is to say, they produce on an average about one-fifth as many children. So only one-fifth of the dwarfs alive at any time are the progeny of dwarfs, the others being the progeny of normal parents. If all dwarfs of this kind were sterilized we could only cut down the number of dwarfs by one-fifth.

With haemophilia we could cut down the frequency to about one-half by preventing the breeding of haemophilics and heterozygous women. With hereditary cataract we could cut down the frequency to much less than one-half—perhaps to one-tenth. Some, though not all, types of mental defect could be considerably reduced; so could harelip and many other physical defects. This would be well worth doing, but the battle would never be finally won, the race never finally purified.

We could, however, cut down the incidence of a great many congenital maladies to a large extent.   Others, such as neonatal jaundice, or *erythroblastosis fetalis,* and perhaps mongoloid idiocy, are due to gene differences between father and mother.   We could only abolish them by forbidding unions between people of different genotypes.   A closed mating system based on skin colour is bad enough, in the sense of making for the division and perhaps the instability of the community.   One based on blood antigens, in which an Rh-negative woman might not marry an Rh-positive man, would perhaps be even worse.   I should be the last to recommend it, even if it saved the lives of a few babies.

Nor at the present time can we do much to diminish the frequency of undesirable recessive conditions, whether they are lethal, like foetal ichthyosis, or merely a slight handicap, like albinism.   The most efficient eugenic method is the introduction of good road transport into backward rural areas, thus encouraging outbreeding.

It may be that, if we knew enough, 1 per cent. or even rather more of the population would be found to carry undesirable dominants or sex-linked recessives, which any sound eugenic policy would reduce.   How should we do it?   Many people believe that carriers should be sterilized, either voluntarily as in Denmark, or compulsorily, as in Nazi Germany.   I do not, for the following reasons.   Laws for compulsory sterilization are liable to gross abuse.   Those for voluntary sterilization are only rather less so.   I recall the case of a labourer in one of your Western states who was given an indeterminate sentence up to five years' imprisonment for theft.   The judge suggested that he be voluntarily sterilized.   He agreed, and was not imprisoned. His agreement can only be called voluntary in a Pickwickian sense.   However, I might be in favour of sterilization if it would finally rid us of these undesirable genes.   But it would not.

There is another reason, perhaps worthy of consideration. If we are ever to control our evolution we shall certainly have to overhaul our whole mating system.   By this I do not mean that we shall have to abolish marriage or adopt polygamy. I do not know what we shall have to do. We shall only do what is right if people realize that we have a duty to beget and bear the best-endowed children possible.

It is of the utmost importance that the idea should not be spread abroad that we can improve the human race to any serious extent by sterilizing individuals who do not come up to certain standards. In England we are already beginning to persuade people with harmful dominants to refrain from reproduction, either by chastity or by contraception. We shall not improve the human race by compulsion. A prerequisite for doing so is the moralization of our sexual behaviour—that is to say, making it subordinate to ideal ends, not to impulse on the one hand or superstition on the other.

DIFFICULTIES OF POSITIVE EUGENICS

What about positive eugenics? In many human societies those types which are most admired are bred out. The Middle Ages admired holiness and courage. The holy men and women were celibate, the courageous men killed one another. Our age admires money-making. The men who make most money have least children. I am less worried by this than many of my contemporaries. I am not convinced that a business executive is a higher type than a saint or even a feudal knight. In any case, a differential birth rate lasting over a century need no more permanently affect the gene frequencies of a race than selection of certain chromosome orders for a few months per year affects a *Drosophila* species. And in Sweden the tendency has already reversed itself, and the poor breed rather more slowly than the rich.

Why, it may be asked, should we not encourage the breeding of rare and desirable genes as we can discourage the breeding of rare and undesirable ones? The answer is that we do not know of a single rare gene in man whose frequency we should increase. I have no doubt that such exist. But our analysis of the genetic basis of human abilities is so utterly rudimentary that we know nothing of them. Their discovery will need a vast programme of collaboration between geneticists, physiologists, and psychologists. Until even one such gene is known, it seems to me rather futile to talk about a programme for positive eugenics.

I would, however, suggest that among the genes whose spread we would want to encourage are those for the non-development of teeth, particularly wisdom teeth. Our cerebral development has caused a good deal of overcrowding of our teeth. I hope also that we shall do something about our noses, which are one of our weak points. (I have a nasal infection at the moment. No other organ lets me down so frequently.) The nose has of course been squashed out of shape by the growth of the brain. In consequence, while a sneeze takes a straight path in a dog or a horse, it has to take a hairpin bend in our own species. In a century or so we may know of detailed changes in our psychological make-up which are equally desirable.

In fact, while we can begin with negative eugenics, we cannot begin on positive eugenics until we have got a great deal more knowledge and a wider diffusion of the eugenic attitude. Probably the first requisites for the development of this knowledge are, on the one hand, the mapping of the human chromosomes, a task to which I have devoted some effort, and on the other, an attempt to analyze the psychological make-up of people judged to be exceptionally gifted, as Spearman in England and the Chicago school in America have tried to analyze that of more normal people. When these are accomplished it will be time to start research on the genetics of great intellectual or moral endowment. Much of it may turn out to be due to heterosis, and as unfixable as the good points of a mongrel dog, but I have little doubt that many rare and desirable genes for these characters exist.

So far I have assumed that our descendants will take over the control of evolution in an intelligent manner. Let us consider the other possibilities. In the next century the human race may largely destroy itself. From the genetic point of view a war using atomic energy would be worse than one using old-fashioned weapons, or even pestilences. For the survivors of Hiroshima and Nagasaki have been so affected that their descendants will show a variety of abnormalities. Some will appear in the first generation and disappear within ten or so.   Others will be recessive, and first appear after several generations, their evil (and very rarely good) effects continuing for many thousand generations. The killing of 10 per cent. of civilized humanity by atomic bombs might not end civilization. The vast crop of abnormalities produced by another irradiated 10 per cent. might do so, and even render recovery very difficult.*

I hope that we shall avoid such an international war, or, what seems to me just as likely, a civil war in which a small group get control of some atomic bombs and hold up a whole nation.   If so, we may settle down to some peaceful world order, but do nothing about our evolution.   In such a case we might stay put for a very long time.   Sewall Wright has shown that, on certain assumptions, which seem to me thoroughly sound, evolution goes on quickest in a species divided up into many groups of a few score or hundred individuals nearly, but not quite, sexually isolated from other groups.   This was the human condition for thousands of centuries during the Old Stone Age.   With agriculture and industry the community has grown, and evolution has probably slowed down.   For some time there was heavy selection against crowd diseases, but the progress of hygiene has checked this tendency.

I do not know what we are selecting for now.   Let me take an example.   Until two generations ago large families were respectable in my country.   Anyone who voluntarily restricted his or her family was a deviant.   Selection favoured genes making for conformity to mores in this respect.   Now it is a deviation from the norm to have a dozen children.   We are selecting in favour of deviation, instead of against it.

We may be favouring genes which make for high sexual activity, low intelligence, or lack of susceptibility to propaganda, to mention only three possibilities.   Most eugenists regard the parents of large families as, on the whole, undesirable genetically. This may well be true.   It is certain that on the whole they are economically unsuccessful.   Before we equate economic success and long term biological value, however, it might be desirable to read the Sermon on the Mount or the record of the dinosaurs.

* Recent calculations suggest that this danger has been exaggerated.

I do not know if the trend described is desirable or not, and I contend that no one else does.

Another possibility is that we shall control our evolution and choose the wrong path. If I had had to pick hopeful ancestors for a rational and skilful animal from past faunas I doubt if I should ever have got the right answer between the Pennsylvanian and the Miocene. I should certainly have picked *Struthiomimus,* a Cretaceous reptile like an ostrich, standing on its hind legs, but with arms in place of wings. I am equally sure that I should go wrong to-day. Dr. H. J. Muller has suggested a method for the radical improvement of the human race, involving the widespread use of artificial insemination. I guess that if I were made eugenic world dictator I should have one chance in a hundred of choosing the right path. Dr. Muller is ten times as good a geneticist as I, so he might have one chance in ten, but not, I think, much more.

I am convinced that the knowledge required, both of past evolution and present genetics and cytology, is considerably greater than the whole body of scientific knowledge on which our present civilization is based. We can get this knowledge if we want it. We may say that God is now enlarging the sphere of human choice, and therefore giving us new duties. Or we may say that the evolutionary process is now passing from the stage of unconsciousness to that of consciousness. But we have not yet got the knowledge.

Our immediate task is the remodelling of human society. This can be done in a few generations. The great men who founded your Republic based it on principles which had never before been applied to human societies but which nevertheless worked in practice. The great men behind the French and Russian revolutions made somewhat different but comparable experiments. No society is perfect, and the time scale of social change is so vastly less than that of evolutionary change that the duty to reform society is far more urgent than that to control evolution. The two duties must and will go together. But it would be fatal to think of the man of the future as one who would fit into contemporary American, British, Russian, or Chinese society, or into any society which we can even imagine to-day.

If I am right he would probably be regarded as a physical, mental, and moral defective. As an adult he would probably have great muscular skill but little muscular strength, a large head, fewer teeth than we have, and so on. He would develop very slowly, perhaps not learning to speak till 5 years of age, but continuing to learn up to maturity at the age of 40, and then living for several centuries. He would be more rational and less instinctive than we are, less subject to sexual and parental emotions, to rage on the one hand and so-called herd instinct on the other. His motivation would depend far more than ours on education. In his own society he would be a good citizen, in ours perhaps a criminal or a lunatic. He would be of high general intelligence by our standards, and most individuals would have some special aptitude developed to the degree which we call genius.

But just as, were we transported to the past, we should be unlikely to win the admiration of *Sinanthropus*, so, were one of these products of planned evolution brought back to our own time, we should probably judge him an unpleasant individual. this thought need not distress us.   We shall not meet him.

INDEX

THE   END